Contributions to Statistics

V. Fedorov / W. G. Müller / I. N. Vuchkov (Eds.) Model-Oriented Data Analysis, XII/248 pages. 1992

J. Antoch (Ed.) Computational Aspects of Model Choice, VII / 285 pages, 1993

W. G. Müller / H. P. Wynn / A. A. Zhigljavsky (Eds.) Model-Oriented Data Analysis XIII/287 pages, 1993

P. Mandl / M. Hušková (Eds.) Asymptotic Statistics X/474 pages, 1994

P. Dirschedl / R. Ostermann (Eds.) Computational Statistics VII/553 Pages, 1994

C. P. Kitsos / W. G. Müller (Eds.) MODA4 - Advances in Model-Oriented Data Analysis XIV / 297 pages, 1995

Heinz Schmidli

Reduced Rank Regression

With Applications to Quantitative Structure-Activity Relationships

With 31 Figures

Physica-Verlag
A Springer-Verlag Company

Series Editors
Werner A. Müller
Peter Schuster

Author
Dr. Heinz Schmidli
Mathematical Applications
CIBA-GEIGY Ltd.
CH-4002 Basel, Switzerland

ISBN 978-3-7908-0871-1 ISBN 978-3-642-50015-2 (eBook)
DOI 10.1007/978-3-642-50015-2

CIP-Titelaufnahme der Deutschen Bibliothek
Schmidli, Heinz:
Reduced rank regression: with applications to quantitative structure activity relationships / Heinz Schmidli. - Heidelberg: Physica-Verl., 1995
(Contributions to statistics)

88/2202-543210 - Printed on acid-free paper

Preface

Reduced rank regression models, introduced by Anderson and others in the 1940's, have experienced a revived interest in recent years. In econometrics, they are now used for the analysis of multivariate time series in connection with the cointegration theory as well as for the study of simultaneous equations systems. Further, in psychometrics, reduced rank regression models have greatly helped to better understand traditional methods such as redundancy analysis and have been important for the development of more adequate graphical methods such as biplots. Whereas in chemometrics, reduced rank regression models are beginning to be used as alternatives to partial least squares for multivariate prediction.

This research monograph is aimed at statisticians interested in this increasingly important topic in multivariate analysis. A basic knowledge of multivariate statistics and matrix algebra is however assumed as a prerequisite.

Chapters 4 and 5 make up the core of the book, in which both classical and Bayesian approaches to reduced rank regression are presented. Chapter 2 gives a general introduction to multivariate prediction and at least section 2.3 should be read, where the prediction criteria are defined. Chapter 3, which discusses two heuristic prediction methods, principal component regression and partial least squares, can be skipped by those mainly interested in reduced rank regression.

The monograph is illustrated and motivated by a single important application area, namely the field of quantitative structure activity relationships (QSAR), introduced in chapter 1. People not interested in this area may omit this chapter, perhaps with the exception of section 1.5.

The algorithms used for reduced rank regression, principal component regression and partial least squares are described in the appendix and are easily implemented in any matrix language such as S–Plus or SAS/IML. A list of the mainly used notations and abbreviations is provided at the end of the book, as well as a subject index.

This monograph is based upon a dissertation presented for a doctorate at the University of Basel. I would like to thank my supervisor Prof. Wolfgang Polasek for his assistance throughout the evolution of this thesis, as well as Prof. Claus Weihs and Prof. Adrian F.M. Smith for many stimulating discussions. I would also like to thank Prof. Helge Toutenburg for helpful comments and additional references, and especially Dr. Hugo Flühler and my colleagues in Mathematical Applications at CIBA–GEIGY Ltd. for their support and encouragement. Finally, thanks are also due to those chemists and biologists of CIBA–GEIGY Ltd. who provided the data and helpful information on QSAR studies.

Basel, April 1995 Heinz Schmidli

Contents

Introduction

In quantitative science, two main principles apply to any new theory. First, it must be testable, that is a theory must allow to make statements about observable phenomena. Second, it must be parsimonious, that is if another theory gives similar answers to the considered problems, the simpler theory should be preferred. One would therefore think that predictability and parsimony are of fundamental interest if statistical methods are applied to real problems. And in fact, parsimony is an essential issue in applied statistics: variable selection methods are extensively discussed, tests are developed to decide whether a simpler model describes the data as well as a more complex one, and so on. In contrast to parsimony, the vital importance of prediction in applied statistics is much less accepted (see, for example, Aitchison and Dunsmore, 1975). This neglect of the predictivistic approach urged many statisticians to argue that "the inferential emphasis of statistics, theory and concomitant methodology, has been largely misplaced" (Geisser, 1980). Or, even more polemical, that "if statisticians would aspire to scientific respectability, [...] inferences we make should be about real, observable quantities" (Dawid, 1993). While these extreme positions are not completely adopted, "predictability and parsimony" is the central motto of this treatise.

Real problems have often stimulated research in both theoretical and applied statistics. The methods presented in the following are to a large extent motivated by an application area of paramount importance to agrochemical and pharmaceutical industries: the development of new biologically active chemical substances, such as new drugs, herbicides or insecticides. An important part in this process is the search for a simple model which allows the prediction of biological activity by structural properties of the chemical substances, that is the search for quantitative structure–activity relationships (QSAR). For a short overview, see chapter 1.

In QSAR, both structural properties and biological activity are described by a multivariate variable. Hence, the main scientific problem may be stated more generally as the search for a parsimonious model which allows the prediction of a multivariate response by a multivariate predictor. Such a task is not only relevant for QSAR studies but for many applications in the natural and social sciences, for example in econometrics where exogenous and endogenous variables are related.

In chapter 2, a general overview of multivariate prediction methods is given. To assess the predictive power of such prediction models, both numerical criteria and exploratory graphical methods are discussed. Prediction criteria such as the mean squared error of prediction and the information criterion are estimated by asymptotic approximations and by crossvalidation (Stone, 1974), a resampling method particularly attractive in a prediction context. Finally, the use of the numerical criteria and the graphical methods for finding a parsimonious and well predicting method is shown.

Two heuristic multivariate prediction methods, principal component regression (PCR, Massy, 1965) and partial least squares (PLS, Wold, 1982), are presented in chapter 3 as a prelude to the mainly considered class of prediction methods, the reduced rank regression models. Both PCR and PLS are methods which have been developed in an ad–hoc manner, but which proved to have good predictive power in many simulation studies and real applications, notably in the area of chemometrics. In this monograph, they are considered as reference methods to the more elaborate reduced rank regression models presented in chapters 4 and 5.

Multivariate regression is widely used to relate a set of response variables to a set of predictor variables. This model assumes that the expected values of the responses are linearly linked with the predictors by an unknown arbitrary coefficient matrix. In reduced rank regression (Anderson, 1951) this coefficient matrix is constrained to be of fixed known rank. It can be shown that the reduced rank regression models are closely related to any standard linear multivariate method such as canonical correlation analysis, redundancy analysis (van den Wollenberg, 1977) or canonical discriminant analysis. Reduced rank regression models have proved to be useful in many applications in the fields of biometrics, chemometrics, econometrics and psychometrics. As explained in section 4.2, they are particularly natural in a QSAR context. To use these models for the prediction of a multivariate response for given multivariate predictor, the unknown parameters in the models have to be estimated. Maximum likelihood estimates can usually be obtained analytically and are then equivalent to weighted least squares solutions. However, for a model with diagonal error covariance matrix, an iterative numerical procedure is necessary to calculate the maximum likelihood estimates; the proposed Gauss–Seidel algorithm is similar to the ECM method (Meng and Rubin, 1993), an extension of the EM algorithm (Dempster et al.,1977). To assess the goodness of prediction of reduced rank regression models, the prediction criteria introduced in chapter 2 are estimated by both asymptotic approximation and crossvalidation. The asymptotic and crossvalidation estimates of the prediction criteria are then used for the selection of the dimensionality of the models, that is the rank of the coefficient matrix, as well as for the selection of the predictor variables. Small simulation studies are intended to give some hints about the appropriateness of the prediction criteria for model selection. Besides these numerical prediction criteria, exploratory graphical methods for model selection are discussed as well.

Predictive inference is often regarded as more natural in a Bayesian setting (see, for example, Aitchison and Dunsmore, 1975). Since the predictive approach is emphasized in this text, a Bayesian analysis of the reduced rank regression models introduced in chapter 4 is presented in chapter 5. Until recently, a Bayesian analysis of reduced rank regression models would have been intractable due to the complexity of the models. However, the awareness of the power of Markov chain Monte Carlo methods, especially the Gibbs sampler (see, for example, Gelfand and Smith, 1990) has changed this situation. A Bayesian analysis of realistic models can now be sometimes even less difficult to obtain than an analysis by classical statistical methods. A Bayesian reduced rank regression model is defined by the specification of the full conditional posterior distribution of the parameters as the basis for the Gibbs sampler. Particular emphasis is placed on the parametrization of the model which is essential for interpretation. The Gibbs sampling procedure can be used very naturally for reduced rank regression models to derive the posterior and predictive distributions. This predictive distribution is then employed to assess the goodness of prediction as well as the adequacy of the model.

Throughout, all methods are illustrated by a QSAR example introduced in section 1.6, which considers the development of an putative insecticide. In chapter 6, two additional quantitative structure–activity relationships case studies from herbicide and drug development are presented. Finally, a discussion in chapter 7 summarizes the main results.

1.
Quantitative Structure Activity Relationships (QSAR)

1.1. Introduction

The development of new effective chemical substances, such as new drugs, herbicides or insecticides, is essential for the survival of an agrochemical and pharmaceutical industry. The process of finding biologically active compounds is very research–intensive and time consuming, illustrated by the fact that typically only about 1 of 10'000 synthesized chemical substances can finally be used as a commercial product. The starting point for the development of a new substance is typically a putative "lead substance", i.e. a substance believed to be chemically modifiable into a biologically very active compound. Once this lead substance is found, numerous modified substances are synthesized and their biological activity is measured. By relating the structural properties of these substances to their biological activity, the physico–chemical properties necessary for high biological activity are identified. This information is then used to synthesize new substances with hopefully higher biological activity. Then a new cycle in the process of developing new active substances is started.

QSAR (quantitative structure–activity relationships) is a rationalization of part of the process to obtain biologically active substances, once a possible lead substance has been found. In a more strict sense, the goal of QSAR is to derive a quantitative model for the relationship between chemical structure and biological activity, where the structure is characterized by descriptors of the physico–chemical properties of the substances and the biological activity is characterized by indicators of the biological effect of the substances. Since various sources of random error are involved in the measurement of biological activity, statistical methods are used to model the relationship between structure and activity.

An adequate application of statistical methods requires some basic insight into the data generation process; hence in the following sections some of the more relevant aspects of QSAR will be shortly discussed.

A very large number of publications are available on the subject of QSAR. An informal overview can be found, for example, in Nirmalakhandan and Speece (1988), and a discussion of statistical thinking in QSAR is given in DunnIII (1989) and Stone and Jonathan (1993, 1994).

1.2. Modification of Substances

The central step in the development of a new biologically active substance from a lead substance is the synthesis of a group of modifications, often called analogs or derivatives. Typically, a possible lead substance is modified by substitution of a part of its molecular structure by some other molecule called substituent. A detailed example for the generation of derivatives is given in section 1.6.

How to modify the lead substance is mainly based on the expertise of the chemist. He/she will try to synthesize a group of substances with very different physico–chemical properties, for example by using very diverse substituents, since substances with similar physico–chemical properties typically have similar biological activity, and hence do not provide additional information. Even if only simple types of modifications are considered, the number of possible derivable substances is enormous, making the choice a very difficult one. For example, if at each of the 6 positions of a benzene, 10 different substituents could be used, 1 million different substances would be obtained. The ideas of experimental design, while certainly of great usefulness, have only been recently introduced in this area (see, for example, Borth et al.,1985, Wold et al., 1986) and the choice of the substituents is still usually made informally by the chemist.

Another important point that has to be considered is the ease with which the modifications can be synthesized. The synthesis of a new substance is very time consuming and may well take one to three weeks. Hence, once a simple synthesis path to produce a particular modification has been found, some other modifications, using a very similar synthesis path are often produced as well.

As a summary, in practice the substances are usually not optimally selected from a statistical point of view.

1.3. Physico–Chemical Descriptors

Obviously the physico–chemical properties of a substance are the cause for its effect on biological systems. However why a substance has the observed effect is seldom known in advance due to the high complexity of the interactions between substance and biological system. Hence it is generally not known a–priori, which of the vast number of physico–chemical descriptors might be of relevance. Nevertheless, a large number of QSAR studies as well as some insights obtained from biochemistry and biophysics make it plausible that a lipophilicity descriptor and some selected descriptors characterizing electronic and steric characteristics are often able to explain a large amount of the biological variation. In addition, some specific knowledge about the mechanism of the effect of a substance on a biological system may suggest some further more specific physico–chemical descriptors.

For relatively simple types of modification by substitution, large databases of substituent descriptors can be used to describe their physico–chemical properties. Alternatively or additionally, quantum–mechanical methods can be used to calculate physico–chemical descriptors. Sometimes measured physico–chemical properties are available as well. However, it is very advantageous to use physico–chemical descriptors which can be calculated, i.e. without the need to actually synthesize the new chemical substance. This allows the prediction of biological activity for not yet synthesized substances, once a model for the prediction of biological activity from physico–chemical descriptors has been found.

As a summary, the structures of the substances are characterized by a number of physico–chemical descriptors, regarded as potentially important for biological activity. Due to the usually informal choice of the modified substances mentioned above, i.e. the absence of experimental design ideas, these descriptors are often correlated.

1.4. Biological Descriptors

The effect of a substance on a biological system, i.e. its biological activity, has always to be seen in connection with the intended use of the substance. For example, an intended anti–inflammatory drug should reduce inflammation, a potential insecticide kill pest insects, a herbicide damage weeds. If these biological activities have to be quantified, typically a large number of variables are plausible indicators of the biological effect. For example, the damage of a weed could be characterized by the reduction of height growth or of the mean size of the leafs. Additionally, not only one species of weed, but a large number of different weeds could be used. If it is known that a substance is effective by blocking a specific enzyme needed for weed growth, an in–vitro test for measuring the inhibiting potential could be developed as well. As a further complication, the effect of a substance on a biological system depends on the actual concentration of the applied substance. More precisely, the intensity of the reaction of the biological system to the substance is generally a monotonous function of the concentration. It is therefore a standard practice, to measure an indicator of biological activity for a number of concentrations of the substance and then to estimate the response as that concentration, for which the expected value of the indicator is some fixed value. For example, if the indicator of the damage of a weed is the height of the weed, the response could be defined as that concentration which results on average in a height of the weed which is only half the height of an untreated weed. This inhibiting concentration is denoted by IC50.

Since the biological activity can be measured by a large number of different descriptors, typically more than one indicator is used. However, if the mechanism for the effect of the substance on the different indicators of biological activity is essentially the same, it is quite plausible that the indicators of biological activity are closely related. To be more specific, it is then reasonable to assume that at least for some subgroups of the biological descriptors, the indicators of biological activity are monotonous functions of each other. For example, a herbicide might have a very similar effect on two species of weed, but if one weed is more robust than the other, higher concentrations of the substance might be necessary for that weed to obtain the same effect. This important aspect of the indicators of biological activity will be further discussed in section 4.2, where it motivates the use of the so called reduced rank regression models.

As a summary, the biological activity of the substances is characterized by a number of biological descriptors. Due to similar effect mechanisms, these descriptors are often related to each other.

1.5. Prediction Model

The central part of a QSAR is the development of a quantitative prediction model, which allows to predict the biological descriptors by physico–chemical descriptors. The model is estimated based on a learning dataset which consists of a number n of substances, for which both physico–chemical descriptors and biological descriptors are available. More formally, the learning data consists of an $n \times p$ predictor matrix X of p physico–chemical descriptors calculated for each of the n substances, and of an $n \times q$ response matrix Y of q biological descriptors, measured subject to error for each of the n substances (see Figure 1.5.1). In typical QSAR–studies, the number of substances is not very high, starting perhaps with as few as 15 substances. The number of initially used physico–chemical variables is often in the range between 5 and 20. The number of measured response variables depends mainly on the application area; often 2 response variables are available, but occasionally as many as 50 indicators of biological activity are measured.

Figure 1.5.1 *Data structure for QSAR.*

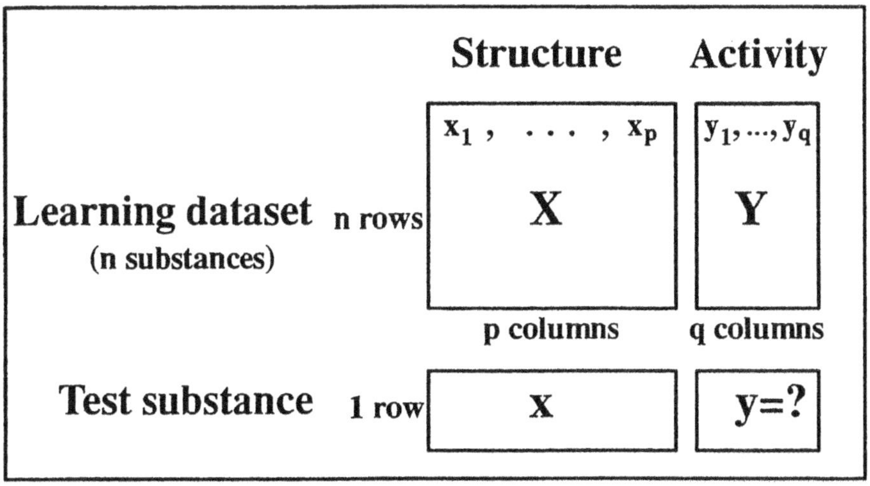

For each substance, the biological descriptors are measured in a separate experiment, so that the errors of the biological descriptors can be assumed to be independent for different substances. This is not always the case for the errors among the biological descriptors, since they are sometimes simultaneously measured in the same experiment.

A prediction model based on the learning data X, Y allows to predict the q–dimensional response vector y that would be measured as the biological effect of a newly synthesized test structure characterized by its p–dimensional vector x of physico–chemical properties. The predicted value is given by $\hat{y} = f(x; X, Y)$, where the function f is chosen by the modeler.

In the field of QSAR, multiple (stepwise) linear regression seems to be predominant (see Rose et al., 1990). The multivariate nature of the biological data is usually not taken into account, although there are noteworthy exceptions (see, for example, Mager, 1984).

1.6. The Development of an Insecticide: an Example

A step in the development of an insecticide in CIBA–GEIGY Ltd. is used to illustrate the preceding discussions. More detailed information about this example can be found in Pascual et al. (1994), from which the following description is taken. A set of 27 substances were synthesized as derivatives of a putative lead structure. The modifications were obtained by substituting the atoms or substructures at 5 different positions of the basic molecular structure shown in Figure 1.6.1, as follows:

- At two positions of one aromatic ring (p2,p3), the H–atom is either replaced by a Cl–atom, or left unchanged;
- At a third position of the aromatic ring (p4), the H–atom is either replaced by one of the substituents Cl, F, CH_3O, t–C_4H_9, CH_3CO, $(CH_3)_2N$, CH_3S, CH_3, C_6H_5CO, or left unchanged;
- At a fourth position (i1), the substructure t–C_4H_9 is either replaced by the substructure i–C_3H_7, or left unchanged;
- At a fifth position (i2), the substructure NH–CS–NH is either replaced by the substructure N=C=N, or left unchanged.

Figure 1.6.1 *The basic structure of the putative lead substance. 5 positions (p2, p3, p4, i1, i2) are modified.*

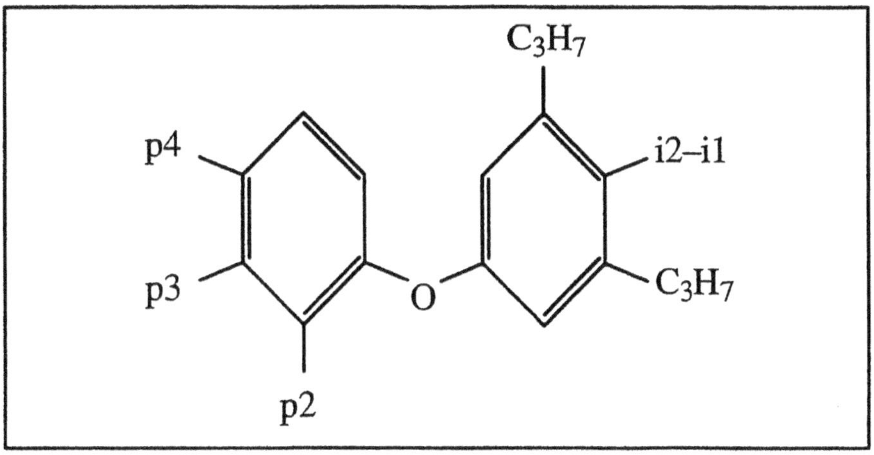

To describe the structure of these substances, the following physico–chemical descriptors are used.

- I1 : indicator variable, indicating if the substructure in fourth position (i1) is replaced (I1=1) or not (I1=0);

- I2 : indicator variable, indicating if the substructure in fifth position (i2) is replaced (I2=1) or not (I2=0);

- IP23 : indicator variable, indicating if one of the two H–atoms in positions p2,p3 is replaced by a Cl–atom (IP23=1) or not (IP23=0);

- Pi, F, R, MR, L, B1 : substituent constants characterizing the physico–chemical properties of the substituent replaced in position p4. More precisely, Pi is a measure of the lipophilicity of the substituent, F and R are Swain and Lupton's measures of its electronic properties, and MR, L and B1 describe some of its steric aspects, namely its overall size, its length and its width (see, for example, Waterbeemd and Testa, 1987). Pi, MR, F and R are calculated from Hansch and Leo (1979), L and B1 from Verloop et al. (1976).

Since the substances are intended as insecticides, their biological activity is defined as their ability to kill larvae and adults of a species of pest insects, the carmine spider mite (*Tetranychus cinnabarinus*).

Table 1.6.1 Insecticide Data

Learning Dataset

	p2	p3	p4	ADULTS	LARVAE	I1	I2	IP23	Pi	F	R	MR	L	B1
IIaa	H	H	H	-2.00	-1.70	0	0	0	0.00	0.00	0.00	1.03	2.06	1.00
IIab	H	H	H	-2.60	-2.60	1	0	0	0.00	0.00	0.00	1.03	2.06	1.00
IIba	H	H	Cl	-1.67	-2.60	0	0	0	0.71	0.41	-0.15	6.03	3.52	1.80
IIda	Cl	H	Cl	-2.00	-2.00	0	0	1	0.71	0.41	-0.15	6.03	3.52	1.80
IIdb	Cl	H	Cl	-2.00	-2.00	1	0	1	0.71	0.41	-0.15	6.03	3.52	1.80
IIea	H	Cl	Cl	-1.70	-1.70	0	0	1	0.71	0.41	-0.15	6.03	3.52	1.80
IIeb	H	Cl	Cl	-2.60	-2.60	1	0	1	0.71	0.41	-0.15	6.03	3.52	1.80
IIfa	H	H	CH30	-1.08	-1.08	0	0	0	-0.08	0.26	-0.51	7.87	3.89	1.35
IIga	H	H	t-C4H9	-1.70	-2.00	0	0	0	1.98	-0.07	-0.13	19.62	4.11	2.59
IIha	H	H	CH3CO	-2.30	-2.60	0	0	0	-0.56	0.32	0.20	11.18	4.06	1.90
IIka	H	H	CH3S	-1.70	-1.70	0	0	0	0.56	0.20	-0.18	13.82	4.32	1.70
IIla	H	H	CH3	-1.70	-1.70	0	0	0	0.65	-0.04	-0.13	5.65	3.00	1.52
IIma	H	H	C6H5CO	-1.70	-1.70	0	0	0	1.07	0.30	0.16	30.33	4.57	2.36
IIIab	H	H	H	-1.40	-1.40	1	1	0	0.00	0.00	0.00	1.03	2.06	1.00
IIIba	H	H	Cl	-1.08	-1.70	0	1	0	0.71	0.41	-0.15	6.03	3.52	1.80
IIIbb	H	H	Cl	-1.70	-1.70	1	1	0	0.71	0.41	-0.15	6.03	3.52	1.80
IIIca	H	H	F	-1.40	-1.08	0	1	0	0.14	0.43	-0.34	0.92	2.65	1.35
IIIda	Cl	H	Cl	-1.08	-1.08	0	1	1	0.71	0.41	-0.15	6.03	3.52	1.80
IIIdb	Cl	H	Cl	-2.00	-1.70	1	1	1	0.71	0.41	-0.15	6.03	3.52	1.80
IIIea	H	Cl	Cl	-1.70	-1.70	0	1	1	0.71	0.41	-0.15	6.03	3.52	1.80

IIIeb	H	Cl	Cl	-1.70	-1.70	1	1	1	0.71	0.41	-0.15	6.03	3.52	1.80
IIIfa	H	H	CH3O	-0.78	-1.08	0	1	0	-0.08	0.26	-0.51	7.87	3.89	1.35
IIIfb	H	H	CH3O	-1.70	-1.70	1	1	0	-0.08	0.26	-0.51	7.87	3.89	1.35
IIIga	H	H	t-C4H9	-0.78	-1.08	0	1	0	1.98	-0.07	-0.13	19.62	4.11	2.59
IIIgb	H	H	t-C4H9	-1.40	-1.08	1	1	0	1.98	-0.07	-0.13	19.62	4.11	2.59
IIIha	H	H	CH3CO	-1.40	-1.70	0	1	0	-0.56	0.32	0.20	11.18	4.06	1.90
IIIhb	H	H	CH3CO	-2.60	-2.60	1	1	0	-0.56	0.32	0.20	11.18	4.06	1.90

Test Dataset

	p2	p3	p4	ADULTS	LARVAE	I1	I2	IP23	Pi	F	R	MR	L	B1
IIca	H	H	F	-2.0	>-1.7	0	0	0	0.14	0.43	-0.34	0.92	2.65	1.35
IIcb	H	H	F	>-1.7	-2.0	1	0	0	0.14	0.43	-0.34	0.92	2.65	1.35
IIia	H	H	(CH3)2N	-2.6	-2.6	0	0	0	0.20	0.10	-0.92	15.55	3.53	1.50
IIIia	H	H	(CH3)2N	>-1.7	-2.0	0	1	0	0.20	0.10	-0.92	15.55	3.53	1.50
IIIka	H	H	CH3S	>-1.7	-1.7	0	1	0	0.56	0.20	-0.18	13.82	4.32	1.70
IIIla	H	H	CH3	>-1.7	>-1.7	0	1	0	0.65	-0.04	-0.13	5.65	3.00	1.52
IIIma	H	H	C6H5CO	>-1.7	>-1.7	0	1	0	1.07	0.30	0.16	30.33	4.57	2.36

ADULTS, LARVAE : biological activity (>-1.7 : incomplete information)

I1,I2,IP23,Pi,F,R,MR,L1,B1:physico-chemical descriptors

I1 = 0 for t-C_4H_9 (a) , = 1 for i-C_3H_7 (b)

I2 = 0 for NH-CS-NH (II) , = 1 for N=C=N (III)

IP23 = 0 for no Cl in position p2 or p3, = 1 otherwise

Pi, F, R, MR, L1, B1 : Properties of substituent in position p4

Substance code based on I2 (II-III), P4 (a-m) and I1 (a-b)

Figure 1.6.2 *Effect of the substances on larvae and adult carmine spider mites (small noise added to prevent overlapping in the figure). A diagonal line is shown as well.*

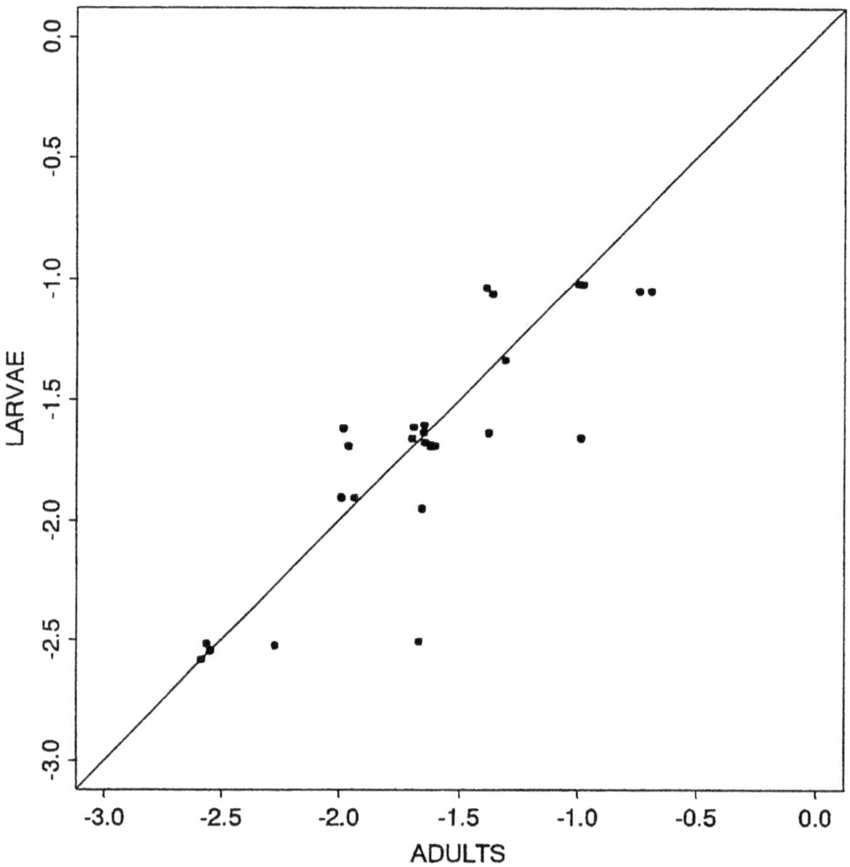

Two response variables are measured for each substance as follows : A number of young bean plants are populated with a mixed population of carmine spider mites. One day later, each of the bean plants is sprayed with an aqueous spray formulation containing different concentrations of the active ingredient. Two days later, the percentage of killed insects is observed. From this data, that concentration of the substance is estimated for which the expected percentage of killed insects is 80%; this concentration is called LD80. The response variable is then defined as log(1/LD80) and high values of this response correspond to high biological activity. The two response variables for larvae and adults are called LARVAE and ADULTS, respectively.

The two response variables are obtained in a common experiment, hence one cannot assume that the errors are independent. Since the biologists are confident that the mechanism causing the insects to die is the same for both larvae and adults, one might think that the response variables are closely related. In fact, a scatterplot of the two variables suggests that approximately a linear relationship exists between the two responses (see Figure 1.6.2).

As a reference, the whole learning dataset can be found in Table 1.6.1 as well as some additionally synthesized test substances with partially incomplete biological information.

2. Linear Multivariate Prediction

2.1. Introduction

In the area of QSAR, described in the preceding chapter 1, a response is measured for a given predictor. There, the main scientific question is how to find a model which allows the prediction of the multivariate response for a given multivariate predictor. The statistician faced with such a prediction problem has to consider many different aspects, for example :

- *Should he/she use flexible nonparametric models or rely on more stable linear procedures?*
  Many very flexible nonparametric nonlinear methods have become available, such as projection pursuit regression (Friedman, 1985), ACE (Breiman and Friedman, 1985), or the methods described in Gifi (1991). All these methods have problems with overfitting, that is they show spurious precision. On the other hand, linear models are generally more stable but introduce severe bias in the case of strong nonlinear relationships. In the QSAR area a relatively small number of observations and a relatively large measurement error of the responses in comparison with their variability is typical. In these circumstances, nonlinear methods can only be used with great care, and linear methods are generally the methods of choice. Hence, nonlinear methods will not be used in the following. Some further discussions about the balance of flexibility and stability, bias and variability can be found in section 4.4.4 with regard to reduced rank regression models.

- *How should a final model be selected among a set of plausible candidates?*
  In applications where knowledge about the relationship between predictors and responses is limited, many prediction methods are often plausible and there is a difficult choice to be made about which prediction method to use. The necessary method selection can either be guided by method selection criteria or by exploratory graphical methods. Since the high number of possible methods generally prohibits a purely exploratory check of models, while the purely criteria–oriented approach might lead to an overlooking of the obvious, a pragmatic strategy can be stated as follows. First use criteria to drastically reduce the number of considered methods, then use exploratory data analysis to select among the reduced set of models the most adequate one. A general discussion of prediction criteria for model selection will be given in section 2.3, the application of these criteria to more specific models in chapters 3 and 4. Exploratory methods for the choice between few prediction methods are presented in section 2.4.

- *Which methods can be used if the predictors are strongly correlated?*
  The presence of multicollinearity among the predictors poses problems to most prediction methods. In a prediction context, this is critical, if the correlation between the predictors is not inherent. To be more precise, suppose that two predictor variables are set in the experiment to values, such that these variables are strongly correlated, while actually these predictor variables can and will be independently varied in the future. Suppose that a response only depends on the first predictor variable. Then no prediction method can decide based on the data which of the two predictor variables affects the response, which then results in unreliable prediction. Additionally, the multicollinearity of the predictor variables can influence the stability of the predictors. To resolve this second issue, two types of approaches are usually taken. One possibility is to reduce the dimension of the predictor space, producing few orthogonal variables containing all relevant information. Examples of such methods are partial least squares (PLS) and principal component regression (PCR) which will be discussed in chapter 3. Another approach to deal with multicollinearity is the method of variable selection, which will be discussed in section 2.5 and in connection with the reduced rank regression models in section 4.6.

- *How do outliers affect the selected method?*
  The prediction methods discussed in chapter 3 and 4 may be influenced by outliers. Robust methods could be used in this case.

However while occasionally mentioning the possibility of robustification (see section 4.4.3), these methods are beyond the scope of this treatise and will not be further discussed.

The basic concepts of multivariate prediction will be presented in the next section 2.2, followed by a detailed discussion of the aspects mentioned above (sections 2.3–2.6). A reference to statistical prediction in general is Aitchison and Dunsmore (1975).

2.2. Multivariate Prediction

Multivariate prediction deals with the simultaneous prediction of more than one response variable by a number of predictor variables. The following general setup of multivariate prediction is used. Suppose that for a fixed predictor variable x_i, a realization of a random variable y_i is observed, $i \in \{1, \dots, n\}$. An important problem of applied statistics is to predict for $k \in \{(n+1),\dots,(n+m)\}$ the distribution of the random variables y_k given x_k (density prediction) or, less ambitious, some of its characteristics, for example the expected value of y_k (point prediction). Obviously, a prediction is only sensible if the distributions of the random variables y_k and y_i are linked. To be more precise, it is assumed that the following model holds.

(2.2.1) $y_i = f(x_i) + e_i \quad , i \in \{1, \dots, (n+m)\}$, where
$f : R^p \rightarrow R^q$ is a vector valued function,
y_i are q–dimensional random variables (response),
x_i are p–dimensional fixed variables (predictor),
e_i are independent identically distributed (i.i.d.) q–dimensional random variables with zero expectation (errors).

It should be noted that the adoption of this model (2.2.1) implies some possibly critical assumptions, namely :

– *independence of the errors* : Independent errors are central for experimentation, and dependencies should usually be regarded as an inappropriateness of the experimental setup. In some QSAR studies, for example if the effect of a herbicide on weed growth is measured in the glasshouse, there might exist a small seasonal effect violating this assumption.

- *identical distribution of the errors (homoscedasticity)* : By a transformation of the responses, it is often possible to approximately fulfill this convenient property. In QSAR studies where often inhibiting concentrations are used as indicators of biological activity, it is standard practice to use the logarithmized values to obtain approximate homoscedasticity.

In contrast to the QSAR area, where these assumptions do not seem to be very critical, there are other application areas, where this is not the case. For example in econometrics, most variables are observed over time, so that errors are often correlated. In this case, generalized regression models should be considered with either arbitrary or structured error covariances (see, for example, Toutenburg, 1992, chapter 4, Toutenburg and Trenkler, 1990).

In the following, occasionally some additional assumptions on the errors are made, for example that the errors are normally distributed. The most drastic assumption that will sometimes be needed concerns the form of the function linking predictors and responses, namely that this relationship is linear. If all these idealized assumptions are combined, the following linear multivariate regression model is obtained.

(2.2.2) $y_i' = \mu' + x_i'M + e_i'$, $i \in \{ 1 , \ldots , (n+m) \}$, where
μ is a q–dimensional vector of unknown parameters (intercept),
M is a $p \times q$ matrix of unknown parameters (coefficient matrix),
e_i are i.i.d. as a multivariate normal $N(0, \Sigma)$ (errors), with
Σ an unknown $q \times q$ error covariance matrix.

For multivariate responses, one might ask why they are not predicted one by one. There are two situations, where this is not adequate :

- *linked errors* : the errors among the responses are correlated, for example if the responses are measured in the same experiment or on the same subject.
- *linked parameters* : the parameters used to model the different responses are related. For example, based on additional knowledge, one can sometimes assume that some parameters are equal (see, for example, Box and Tiao, 1973). If the columns of the coefficient matrix M in (2.2.2) are linearly dependent, a reduced rank regression model is obtained (see chapter 4).

In some applications, both errors and parameters are linked with respect to the responses. If neither errors nor parameters for the different responses are related, then these can and should be modeled independently.

Multivariate regression models have proved to be useful in many other application areas than QSAR. For example in econometrics, multivariate regression models can be derived as the reduced form of a simultaneous equation system. This econometric model links endogenous variables with exogenous variables by assuming that certain linear combinations of both endogenous and exogenous variables are random error variables. For a discussion of this econometric model and the reduced form, see, for example, Toutenburg (1992, pp.3) and section 4.4.5.

In the following, linear multivariate prediction methods will be discussed which are most appropriate for data generated by model (2.2.2), although not too large deviations will not make their application useless. If point prediction methods are discussed, only those will be considered which lead to the following prediction $\hat{y}$ of the expected value of the q–dimensional response vector y for given p–dimensional predictor vector x.

(2.2.3) $\hat{y}' = \hat{\mu}(Y;X)' + x'\hat{M}(Y;X)$, where

$X = (x_1' \dots x_n')'$ is an $n \times p$ known fixed matrix of predictors,

$Y = (y_1' \dots y_n')'$ is an $n \times q$ matrix of observed responses and

$\hat{\mu}(Y;X)', \hat{M}(Y;X)$ are method dependent estimates.

Note that while the prediction $\hat{y}$ is linear in the predictor vector x , the estimate of the coefficient matrix $\hat{M}(Y;X)$ is not assumed to be linear in the response matrix Y , as in ordinary least squares regression. Specific linear prediction methods will be discussed in chapters 3 and 4.

When density predictors are discussed, only those will be considered that can be derived from the point predictors as follows

(2.2.4) $\hat{\rho}_y = N(\hat{y}, \hat{\Sigma})$ (normal density), where

$\hat{\Sigma} = (Y - \hat{\mu} - X\hat{M})'(Y - \hat{\mu} - X\hat{M})/n$ and

$\hat{y}, \hat{\mu}, \hat{M}$ are obtained by (2.2.3) .

Density predictors will be used in the following when the distribution of the q–dimensional response vector y for given p–dimensional predictor vector x is assumed to be given by the model (2.2.2). The density predictor (2.2.4) is then obtained by replacing the unknown parameters by their maximum likelihood estimates.

The prediction rules given by (2.2.3) and (2.2.4) allow to predict the expected value and the density, respectively, of a "future" response y for any vector x in predictor space. In addition to such a prediction rule, one would like to know how good such a prediction rule may be expected to be before any future response is actually observed. In other words, an assessment of the goodness of prediction is needed. Both numerical and graphical methods can be used to judge the predictive power of a prediction method. Numerical prediction criteria will be mainly discussed in section 2.3 and exploratory graphical methods in section 2.4. These numerical and graphical procedures are then used in section 2.5 for method and variables selection, as well as for the assessment of the goodness of prediction of the selected models (section 2.6). The discussion is later taken up again, when these procedures are applied to specific prediction methods, the heuristic multivariate prediction methods in chapter 3 and the reduced rank regression models in chapter 4.

2.3. Prediction Criteria

2.3.1. Introduction

A formal way to assess the goodness of prediction is first to define a loss function to quantify the loss which results from imprecise prediction for each possible case, and then to specify an adequate strategy for an overall assessment. Statistical decision theory can be used to address these issues (see Aitchison and Dunsmore, 1975). A pragmatic procedure is to select both a convenient loss function and a strategy which on the one hand allows some theoretical treatment and on the other hand is reasonably adequate for many real problems. A quadratic loss function is convenient for point predictors while for density predictors the Kullback–Leibler (or Boltzmann) information has some desirable properties (see section 2.3.2). The strategy used in the following is to choose a prediction method with small expected loss (see section 2.3.3). This is the strategy most widely used in the literature and seems to be acceptable in many situations (see, for example, Aitchison and Dunsmore, 1975). Other strategies such as the minimax strategy, which is especially useful when prior information is available (for example if the parameters have to be contained in a bounded set), will not be discussed (see, for example, Toutenburg, 1992, section 3.11).

After the selection of a loss function and strategy, the choice between different predictors would be trivial, if the value of the prediction criteria for each of the considered methods were known. However, since prediction criteria typically depend on unknown parameters, they have to be estimated from the data (see section 2.3.5). In the following, prediction criteria and their estimators will be discussed for both point and density predictors in parallel, in order to emphasize that the same line of reasoning can be used for both types of predictors.

2.3.2. Loss Function

Suppose that for a known fixed $n \times p$ matrix of predictors $X = (x_1' \ldots x_n')'$, an $n \times q$ matrix of responses $Y = (y_1' \ldots y_n')'$ is observed as a realization of a random matrix Y. Similarly, suppose that for a possibly different known fixed $m \times p$ matrix of predictors $\tilde{X} = (\tilde{x}_1' \ldots \tilde{x}_m')'$, a random $m \times q$ matrix $\tilde{Y} = (\tilde{y}_1' \ldots \tilde{y}_m')'$ with density $\rho_{\tilde{Y}}$ is given. A point predictor for the expected value of $\tilde{Y}$ is calculated by (2.2.3) from the data Y and is denoted by $\hat{\tilde{Y}}$. Similarly, a density predictor is calculated by (2.2.4) and denoted by $\hat{\rho}_{\tilde{Y}}$. In the following it is assumed that model (2.2.1) holds for both response matrices.

A loss function has to suitably measure the distance between any realization of the random matrix $\tilde{Y}$ and its prediction. For point predictors, such a measure is the (weighted) squared error of prediction given by

(2.3.1) $1/m \, \| (\tilde{Y} - \hat{\tilde{Y}})\Gamma^{-1/2} \|^2$, where
Γ is a fixed $q \times q$ matrix of weights.

The norm used here and in the following is the Frobenius norm, defined for any matrix A by $\| A \|^2 = \mathrm{tr}A'A$. Obviously, the distance (2.3.1) is minimal, if the predicted value and the realization of the response matrix are identical. A non–trivial issue is the choice of the weighting matrix Γ. One can argue, that an adequate choice for the weighting matrix is the covariance matrix Σ of the measurement error. Indeed, this takes into account that even if the predictor is optimally chosen, any realization of the random matrix $\tilde{Y}$ will deviate from the predicted value due to the measurement error. This choice of the weighting matrix would then measure the goodness of the predictor in the adequate metric. If this error covariance matrix is known to be $\Sigma=\Sigma_0$, then one should choose $\Gamma=\Sigma_0$. If the error covariance matrix is known to be proportional to the identity matrix, then the identity matrix can be selected as the weighting matrix, since any proportionality factor is irrelevant. The situation is more complicated, if the error covariance matrix is either arbitrary positive definite or diagonal. One possibility is to estimate the error covariance matrix from the data and then to replace this estimate as the weighting matrix in (2.3.1). Another possibility is to implicitly assume that the error covariance matrix is proportional to the covariance matrix of the observed responses and then to replace this covariance matrix in (2.3.1) as the weighting matrix. A motivation for this choice of the weighting matrix will be given in connection with the reduced rank regression methods in chapter 4. It should be noted, that in the univariate case, i.e. if there is only one response variable, the choice of the (scalar) weight Γ is irrelevant.

To obtain a distance measure (2.3.1) that is easier to interpret, the performance of the prediction method has to be compared with the performance of the constant predictor, i.e. a procedure which predicts the responses by the mean of the observed responses. A criterion similar to the R^2 criterion of multiple regression is then obtained by choosing Γ to be equal to the diagonal of the covariance matrix of the responses. This then gives the mean of the R^2–values of the individual responses.

(2.3.2) $$R^2_{mean} := 1 - \| (\tilde{Y} - \hat{\tilde{Y}})\Gamma^{-1/2} \|^2 / (q \cdot m) \quad , \text{where}$$
$$\Gamma := \text{diag}(\, (Y - 1/n\; 1_n 1_n'Y)'(Y - 1/n\; 1_n 1_n'Y)/n\,) \;.$$

In the following, when point predictors are considered, this choice of Γ will be used.

For density predictors, a measure for the distance between any realization of the random matrix $\tilde{Y}$ and the predicted density $\hat{\rho}_{\tilde{Y}}$ is the evaluation of the negative logarithm of the predicted density at the realization of the response matrix, and is given by

(2.3.3) $$-\, 1/m \log \hat{\rho}_{\tilde{Y}}(\tilde{Y}) \,.$$

The distance is minimal, if the realization of the response matrix is equal to the modal value of the predicted density. If the predicted density is of the form (2.2.4), this is the case if and only if the realization of the response matrix is identical to the point predictor.

2.3.3. Expected Loss

In the preceding section, a loss has been associated with the deviation of a specific realization of the response from the predictor. This measure of deviation is of limited interest, since one does not know the actual realization in advance. What is useful to know is how a prediction procedure performs on average, i.e. the expected loss. The expected loss is sometimes called risk (compare, for example, Toutenburg, 1992, p.35).

For point prediction the resulting criterion is the (conditional) mean squared error of prediction

$$\text{(2.3.4)} \quad \text{cMSEP} := 1/m \; E_{\tilde{Y}|\tilde{X}} \| (\tilde{Y} - \hat{\tilde{Y}})\Gamma^{-1/2} \|^2 .$$

For density predictors, the (conditional) information criterion is obtained

$$\text{(2.3.5)} \quad \text{cIC} := - \; 1/m \; E_{\tilde{Y}|\tilde{X}} \log(\hat{\rho}_{\tilde{Y}}(\tilde{Y})) .$$

Note that (2.3.5) is closely connected with the Kullback–Leibler (or Boltzmann) information $I(\hat{\rho}_{\tilde{Y}} , \rho_{\tilde{Y}})$, which measures the distance between the predicted density $\hat{\rho}_{\tilde{Y}}$ and the true density $\rho_{\tilde{Y}}$, since

$$\begin{aligned}
\text{(2.3.6)} \quad I(\hat{\rho}_{\tilde{Y}} , \rho_{\tilde{Y}}) &:= \int \log(\rho_{\tilde{Y}}(z) / \hat{\rho}_{\tilde{Y}}(z)) \, \rho_{\tilde{Y}}(z) \, dz \\
&= c - \int \log(\hat{\rho}_{\tilde{Y}}(z)) \, \rho_{\tilde{Y}}(z) \, dz \\
&= c + m \cdot \text{cIC} \qquad \text{, where} \\
c := \int \log(\rho_{\tilde{Y}}(z)) \, \rho_{\tilde{Y}}(z) \, dz & \quad \text{is a constant.}
\end{aligned}$$

For a motivation of this measure in a prediction context, see, for example, Akaike (1974, 1985).

Instead of conditioning on the observed data Y, the unconditional prediction criteria will be used in the following. The mean squared error of prediction is given by

$$\text{(2.3.7)} \quad \text{MSEP} := 1/m \; E_{Y|X} E_{\tilde{Y}|\tilde{X}} \| (\tilde{Y} - \hat{\tilde{Y}})\Gamma^{-1/2} \|^2 .$$

The information criterion is given by

(2.3.8) $\mathrm{IC} := -\ 1/m\ E_{Y|X}E_{\tilde{Y}|\tilde{X}}\ \log(\ \hat{\rho}_{\tilde{Y}}(\tilde{Y})\ )\,.$

2.3.4. Relevant Predictor Space

The goodness of prediction of a method not only depends on the selected loss function and strategy but on the part of the predictor space where the prediction method will be applied. Up to now, the prediction criteria have been defined for general predictor matrix $\tilde{X}$, without being specific. One possibility would be to calculate the prediction criterion for each point of interest in the predictor space separately. With regard to the selection of a prediction rule, this would mean that the choice among the different methods depends on the actually chosen predictor point (see, for example, Allen, 1971). Another possibility would be to specify a density ρ_X on the predictor space giving different importance to different regions and to consider the integrated prediction criterion. While these two approaches have some logical advantages, a much more convenient approach will be taken, namely to consider a replication experiment, i.e. to set the future predictor matrix $\tilde{X}$ equal to the predictor matrix X for which the response matrix Y was observed. This choice has considerable practical advantages, since it allows to compare a predicted value with an already observed response. This is both important for using exploratory graphical methods (see section 2.4), as well as for the estimation of the prediction criteria from the data (see following section 2.3.5).

2.3.5. Estimation of the Expected Loss

The prediction criteria discussed in the previous section depend on the distribution of the responses, which is at least partially unknown. Hence the expected loss has to be estimated. There are three main approaches to estimate the expected loss defined by the mean squared error of prediction MSEP (2.3.7) for point predictors and the information criterion IC (2.3.8) for density predictors, respectively :

- *Fit approach* : Instead of the response matrix $\tilde{Y}$ that would be observed in the future, the actually observed response matrix Y is used.
- *Asymptotic approach* : Large sample approximations of the expected loss are derived.

– *Resampling approach* : Resampling methods are used to obtain approximate estimates of the expected loss.

A general overview of these three estimation procedures will now be given. More specific issues will be discussed when the heuristic multivariate prediction methods (principal component regression and partial least squares, chapter 3) and the reduced rank regression models (chapter 4) are presented.

2.3.5.1. Point Predictors

For point predictors, the mean squared error of prediction MSEP is defined by (2.3.7). In the following it is only assumed that $\tilde{Y}$ and Y follow the very general model (2.2.1) and are independent and identically distributed. Only weak additional assumptions are later needed for the linear multivariate prediction methods. Some insight is gained from decomposing MSEP as follows.

(2.3.9) $$\text{MSEP} = \operatorname{tr}[\, \{\Sigma + \text{bias}'\text{bias} + \text{var} - 2\,\operatorname{cov}(\tilde{Y}, \hat{Y})\,\}\Gamma^{-1}\,],$$

where

$\Sigma := 1/n\, E_{\tilde{Y}|X}(\tilde{Y} - E_{\tilde{Y}|X}\tilde{Y})'(\tilde{Y} - E_{\tilde{Y}|X}\tilde{Y})$ is the error covariance matrix,

$\text{bias} := 1/\sqrt{n}\,(E_{\tilde{Y}|X}\tilde{Y} - E_{Y|X}\hat{Y})$ is the bias of the predictor,

$\text{var} := 1/n\, E_{Y|X}(\hat{Y} - E_{Y|X}\hat{Y})'(\hat{Y} - E_{Y|X}\hat{Y})$ is the average variance of the predictor,

$\operatorname{cov}(\tilde{Y}, \hat{Y}) := 1/n\, E_{\tilde{Y}|X}E_{Y|X}(\tilde{Y} - E_{\tilde{Y}|X}\tilde{Y})'(\hat{Y} - E_{Y|X}\hat{Y})$ is the average covariance between the predicted and the future values.

Note that since $\tilde{Y}$ and Y are assumed to be independent, the last term of (2.3.9), $\operatorname{cov}(\tilde{Y}, \hat{Y})$, vanishes.

A naive way to estimate the MSEP is to replace in (2.3.7) $\tilde{Y}$ by Y, since both are assumed to be identically distributed. However, the decomposition (2.3.9) shows that in this case unbiased estimates of all parts are obtained with the exception of the last term

(2.3.10) $\delta := -2\ \mathrm{tr}[\mathrm{cov}(Y, \hat{Y})\Gamma^{-1}]$

Since this term is usually negative, the MSEP is systematically underestimated by the fit criterion, defined as

(2.3.11) $\mathrm{MSEP_{FIT}} := 1/n \, \| (Y - \hat{\hat{Y}})\Gamma^{-1/2} \|^2$.

Nevertheless, the fit criterion is useful as a lower bound. If for a given learning dataset the $\mathrm{MSEP_{FIT}}$ of a method is too small to be useful, the corresponding method does not have to be further considered. If this is the case for all used methods, this indicates that some important predictor variables have not been included, that there are severe nonlinearities, or that the measurement error is very large.

An alternative approach is to correct the fit estimate $\mathrm{MSEP_{FIT}}$ by estimating δ in (3.1.10) by $\hat{\delta}$ and then use the corrected version

(2.3.12) $\mathrm{MSEP_{ASY}} := \mathrm{MSEP_{FIT}} - \hat{\delta}$.

An estimate can be obtained as follows. First, one can decompose

$Y = Y_0 + E$, where

$Y_0 := E_{Y|X}Y$ and $\mathrm{cov}(\mathrm{vec}(E)) = (\Sigma \otimes I_n)$.

Writing $\hat{\hat{Y}} = X\hat{M}(Y)$, the Taylor expansion of $\hat{M}(Y)$ is assumed to exist and is given by

$\mathrm{vec}(\hat{M}(Y)) = \mathrm{vec}(\hat{M}(Y_0)) + C(Y_0)\mathrm{vec}(E) + r(E)$

for some matrix function $C(Y_0)$, where

$r(E)/ \| E \| \to 0$ as $E \to 0$.

Neglecting $r(E)$, one has

$$\mathrm{tr}(Y - E_{Y|X}Y)'(\hat{\hat{Y}} - E_{Y|X}\hat{\hat{Y}}) = \mathrm{vec}(Y - E_{Y|X}Y)'\mathrm{vec}(\hat{\hat{Y}} - E_{Y|X}\hat{\hat{Y}})$$

$$\approx \mathrm{vec}(E)'(I_q \otimes X)C(Y_0)\mathrm{vec}(E) = \mathrm{tr}[(I_q \otimes X)C(Y_0)\mathrm{vec}(E)\mathrm{vec}(E)'] .$$

Taking the expected value, one finally obtains

$$\text{(2.3.13)} \quad -1/2\ \delta \approx \mathrm{tr}[(I_q \otimes X)C(Y_0)(\Sigma \otimes I_n)(\Gamma^{-1} \otimes I_n)] = \mathrm{tr}[(\Sigma\Gamma^{-1} \otimes X)C(Y_0)] .$$

Note again that no particular model for Y has been assumed; in particular it is not necessary that Y_0 is linear in X.

To illustrate this result, the ordinary least squares estimator is considered with

$$\hat{M}(Y) = PY_0 + PE\,, \text{ where } P = (X'X)^{-1}X'\,.$$

Obviously, the first order Taylor approximation is exact with

$$C(Y_0) = (I_q \otimes P).$$

Hence one has

$$-1/2\delta = \mathrm{tr}[(\Sigma\Gamma^{-1} \otimes X(X'X)^{-1}X')] = p \cdot \mathrm{tr}(\Sigma\Gamma^{-1})\,.$$

Note that this is essentially a generalization of the Cp criterion introduced by Mallows (1973) for multiple linear regression to multivariate regression.

The same idea can be used for the dimension selection in principal component regression (see section 3.4), where again the Taylor approximation is exact.

To obtain an operational procedure, the unknown parameter Σ and possibly the parameters needed to evaluate $C(Y_0)$ have to be estimated. If repeated measurements are available, then Σ can be estimated without any further assumptions. If repeated measurements are not available, then an estimate can be obtained from the fit residuals, assuming that Y_0 is linear in X.

For ordinary least squares regression and for principal component regression (see section 3.4), $C(Y_0)$ does not depend on unknown parameters. However, this is a special case and for the reduced rank regression models introduced in chapter 4, $C(Y_0)$ has to be estimated (see section 4.5.3.1). The asymptotic estimate of the MSEP is then given by

(2.3.14) $$MSEP_{ASY} := MSEP_{FIT} + 2 \cdot \mathrm{tr}[(\hat{\Sigma}\Gamma^{-1} \otimes X)\hat{C}(Y_0)]\,.$$

A third approach is based on the crossvalidation idea (see, for example, Allen, 1974, Stone, 1974, Geisser, 1975), which mimics true prediction by eliminating an observation, calculating a prediction rule based on the remaining data and predicting the response for the left–out observation. If this procedure is repeated for all observations, a crossvalidation prediction matrix $\hat{\hat{Y}}_{(.)}$ is obtained, which can be compared with the observed values. More formally, define $X_{(i)}$ as the matrix obtained from X by the deletion of the i–th row, and similarly $Y_{(i)}$. Then the i–th row of the crossvalidation prediction matrix is defined as

$$\hat{\hat{Y}}_{i.} := X_{i.}\hat{M}(X_{(i)}, Y_{(i)}).$$

The crossvalidation estimate of the MSEP is given by

$$(2.3.15)\quad MSEP_{CV} := 1/n\ \mathrm{tr}[\ (Y - \hat{\hat{Y}}_{(.)})'(Y - \hat{\hat{Y}}_{(.)})\ \Gamma^{-1}\ ]\,.$$

Note again that no particular model for the response matrix has to be assumed.

Alternative resampling methods like p%–crossvalidation and bootstrap methods could be used as well and have been extensively discussed in the literature (see, for example, Efron, 1983, Bunke and Droge, 1984, Weihs and Schmidli, 1991, Shao, 1993).

2.3.5.2. Density Predictors

The information criterion IC is defined by (2.3.8). As for the MSEP, an obvious approach to obtain an estimate of the information criterion is to replace in (2.3.8) $\tilde{Y}$ by Y, since both are assumed to be identically distributed. This gives the fit estimate

$$(2.3.16)\quad IC_{FIT} := -\ 1/n\ \ \log(\ \hat{\rho}_{\tilde{Y}}(Y)\ )\,.$$

However, very similar problems as with the MSEP are encountered. Since the density predictor is of the form (2.2.4), the IC can be written as

$$(2.3.17)\quad IC = 1/2\ E_{Y|X}E_{\tilde{Y}|\tilde{X}}\ (\ \log(\hat{\Sigma})\ +\ \|\,(\tilde{Y} - \hat{\hat{Y}})\hat{\Sigma}^{-1/2}\,\|^2\ )\,.$$

Replacing $\tilde{Y}$ by Y, one has : $\|\,(Y - \hat{\hat{Y}})\hat{\Sigma}^{-1/2}\,\|^2\ =\ nq$,

which usually underestimates the second term of IC in (2.3.17) as seen by a similar argument as the one used for the fit estimate of the MSEP.

As shown by Akaike (1974), an asymptotic correction of the fit estimate of the information criterion can be obtained if a parametric likelihood model for the responses is available. The estimate, usually denoted by AIC, is given by

$$(2.3.18)\quad IC_{ASY} := -\ 1/n\ \log(\ \hat{\rho}_{\tilde{Y}}(Y)\ ) + m/n\ ,\ \text{where}$$

m is the number of free parameters in the prediction model.

This estimate will be used in section 4.5.2.1 for the rank selection of reduced rank regression models, as well as the third estimate of the information criterion , which is obtained by using crossvalidation. The estimate is given by

$$(2.3.19)\quad IC_{CV} := -1/n \sum_{i=1}^{n} \log(\hat{\rho}_{(i)}(Y_{i.}))\ , \text{ with}$$

$$\hat{\rho}_{(i)} = N(\hat{\bar{y}}_{(i)}, \hat{\Sigma}_{(i)})\ , \text{ where } \hat{\bar{y}}_{(i)} \text{ and } \hat{\Sigma}_{(i)} \text{ are obtained from}$$

(2.2.4) using the reduced data $Y_{(i)}, X_{(i)}$ instead of Y, X.

Only the asymptotic estimate of the information criterion is usually discussed in the literature. The exception is Stone (1977) who showed the asymptotic equivalence of (2.3.18) and (2.3.19).

2.4. Exploratory Graphical Methods

Exploratory data analysis (see, for example, Polasek, 1988) plays an important part in applied statistics. A fairly general strategy for the exploratory analysis of multivariate data is presented in Weihs and Schmidli (1990,1991), Schmidli (1991,1992) and Weihs (1993) under the name OMEGA :

"The term Online Multivariate Exploratory Graphical Analysis (OMEGA) describes an online study of relationships in a multivariate data set, where, rather than testing one specific property, as many clues as possible for interesting structures are searched for by different dimension reductions and succeeding graphical analysis. At least three characteristic aspects of such an OMEGA should be emphasized :

- the *experimental character* of the search, in which the data are analysed by interactively (online) switching between different approaches;
- the *dimension reduction*, i.e. the transition from high–dimensional data to a low–dimensional transformation which represents the properties of interest of the data as well as possible;
- the interactive, dynamical *graphical analysis* of the low–dimensional data, exploiting the great perspective power of the human eye for the identification of structure." (Weihs and Schmidli, 1990).

The different methods are organized in the so-called OMEGA-pipeline, corresponding to a sensible structuring of the data analysis (see Figure 2.4.1).

Figure 2.4.1 *The OMEGA pipeline.*

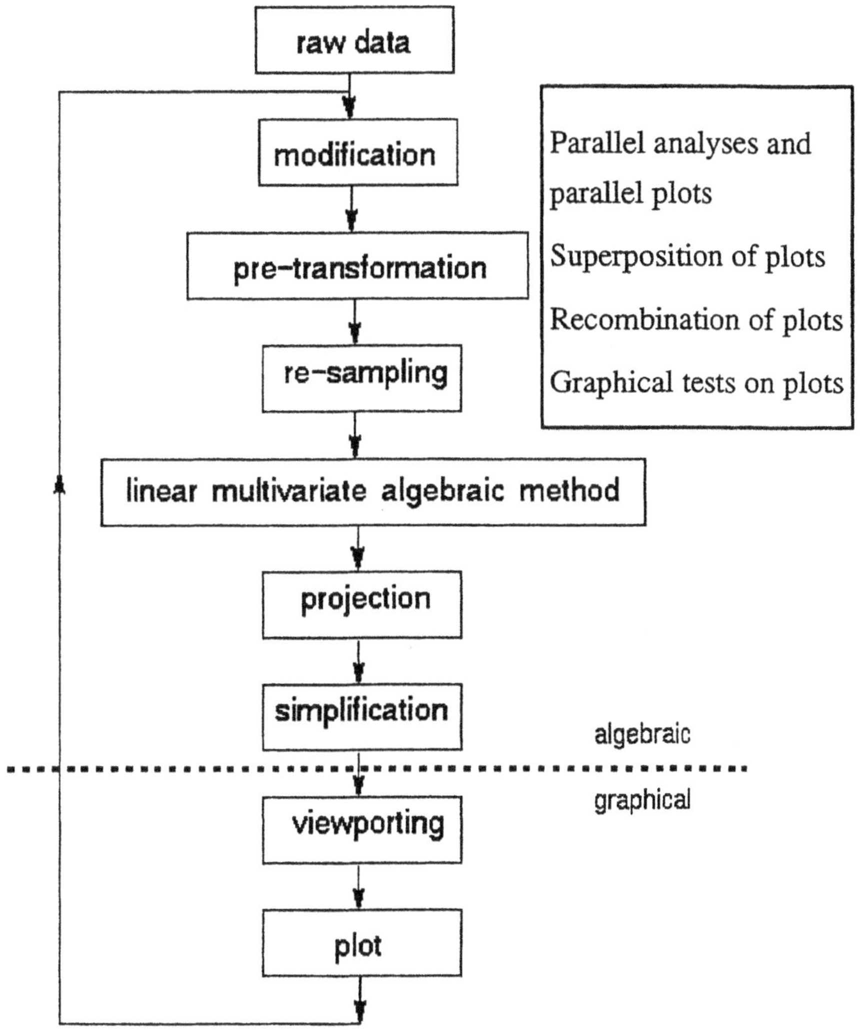

The raw data are modified by selecting a subset of the variables and/or observations. Each variable can then be pre–transformed, for example by centering and/or scaling, or by using a logarithmic transformation. If resampling is used, observations are sampled from this modified and pre–transformed dataset, either by bootstrapping or by systematic elimination of observations (for crossvalidation). Then a linear multivariate algebraic method is applied, for example principal component analysis, which gives new variables, the latent variables or scores, as linear combinations of the original ones. Then a projection takes place by considering only the first few most important latent variables. The latent variables are difficult to interpret, since they usually are linear combinations of all original variables. A simplification finds for each dimension a linear combination of few original variables which is close to the latent variable. These simplified latent variables are then viewported and plotted. Parallel analyses are obtained, for example, by using different linear multivariate methods, different samples in resampling or different pre–transformations. The plots obtained from parallel analyses can be superimposed or recombined to detect more easily similarities and differences. Graphical test can be used, for example, to informally check independence of variables.

OMEGA integrates the algebraic dimension–reduction and the graphical techniques and emphasizes that the same data analysis flow can be used for each of the algebraic methods.

An implementation of the OMEGA concept is described in Schmidli and Weihs (1994). The user interface for this software is written in SAS (SAS Institute Inc., 1990), taking advantage of the interactive graphical methods supplied by SAS/INSIGHT. To allow an interactive analysis, the computer–intensive resampling methods should not take to much time. Hence, the algebraic and resampling methods are written in FORTRAN, heavily relying on the subroutines provided by NAG (NAG Inc., 1990). The OMEGA software runs on UNIX workstations and on PC's.

In a prediction context, the methods of exploratory data analysis as implemented in OMEGA can be used to check the adequacy of a selected prediction method. This mainly consists of a comparison of predicted and observed responses and the analysis of residuals, possibly derived from a crossvalidation procedure. Other aspects such as leverage, individual variable diagnostics and transformation analysis are other important parts of an exploratory analysis. These regression diagnostic procedures are quite far developed for univariate problems, especially for linear regression (see, for example, Belsley et al., 1980), Cook and Weisberg, 1982). Much less attention has been given to the case of multivariate responses. Most diagnostic tools can be applied or slightly adapted for the individual analysis of each response variable. Truly multivariate analysis methods are described, for example, in Gnanadesikan (1977, pp.258), Seber (1984, section 8.5) and Gnanadesikan and Kettenring (1972), and mainly consist of performing a principal component analysis of the residuals.

When several plausible prediction methods are considered, with the goal of selecting the best among them, some additional exploratory methods can be used for comparison purposes, for example by appropriately linking the diagnostic plots obtained from the different methods (see, for example, Stuetzle, 1988).

In the following, only the exploratory graphical methods in OMEGA will be used. Two types of graphical displays will be extensively utilized to compare several prediction methods : the parallel display of observed and predicted responses obtained by different prediction methods and the interpolation of residuals for different prediction methods. These graphical displays are explained when they are first used in section 3.5.

2.5. Method and Variable Selection

2.5.1. Introduction

In applications with only very limited knowledge about the relationship between predictors and responses, many prediction methods are often plausible, so that one has to decide which prediction rule to choose ultimately. Even if an adequate particular multivariate prediction method has been selected, a problem common to all multivariate methods is the choice of its dimensionality. If the appropriate method and dimension have finally be selected, it is often useful both from an interpretational and predictive point of view, to eliminate some of the non–informative predictor variables. Both formal prediction criteria (see section 2.3) and informal exploratory methods (see section 2.4) can be combined to obtain a suitable strategy for the whole process of selecting an appropriate prediction rule. A useful strategy is to rely on prediction criteria for the reduction of the number of plausible methods, and then to compare the most promising candidates by exploratory methods.

For model selection, the fit estimate of the expected loss (see section 2.3.5) is not appropriate, since then typically the biggest model is selected in a series of nested models. For heuristically motivated multivariate prediction methods such as those discussed in chapter 3 (principal component regression and partial least squares), there is often no alternative to resampling methods such as crossvalidation (see section 2.3.5). For methods based on models, such as the reduced rank regression models (see chapter 4), the use of asymptotic estimates can be considered as well (see section 2.3.5).

The predictive approach to method selection is intensively discussed in the literature, especially in connection with resampling procedures such as crossvalidation (see, for example, Allen, 1974, Geisser and Eddy, 1979, Efron, 1983, Weihs and Schmidli, 1991, Shao, 1993). Besides these approaches to the selection of the appropriate prediction model, there are theoretical results which allow to compare prediction methods as well, for example based on the mean squared error of prediction matrix (see, for example, Toutenburg and Trenkler, 1990). Unfortunately, theoretical results are very difficult to obtain if one leaves the class of linear predictors, that is if one considers predictors that are not linear in the responses. Since the methods discussed in the following chapters 3 and 4 (partial least squares and reduced rank regression models) are nonlinear functions of the responses, a theoretical approach along these lines seems not to be possible and hence will not be considered in the following chapters.

2.5.2. Method Selection

If several prediction methods are considered, the choice among them can be based on the prediction criteria discussed in the sections 2.3, and the method with minimal estimated expected loss is then selected. For all the linear multivariate prediction methods considered in chapters 3 and 4, prediction rules are specified once the dimensionality has been selected. Again, for each method that dimension is selected with minimal estimated expected loss.

The diagnostic methods, discussed in section 2.4 are an alternative to criteria based selection procedures, especially if for a given method, prediction rules corresponding to some few dimension reductions are compared.

In practice, especially in the area of quantitative structure–activity relationships (see chapter 1), one will often be able to identify the a–priori most appropriate linear method and sometimes even the dimensionality. In this case, the calculation of the estimated expected loss for other methods and dimensions can be used as a safeguard against the selection of an unexpectedly bad prediction rule.

2.5.3. Variable Selection

Usually predictor variables are included since they possibly might have an effect on the responses. However, many of these predictors turn often out to be useless for prediction, so that one should consider the elimination of these noninformative variables. There are many reasons not to use a larger than necessary number of variables (see, for example, the discussion in Cox and Snell, 1974, Schmidli, 1992).

- *Prediction stability* : The variance of the predictor is often proportional to the number of included variables (for example for the reduced rank regression methods discussed in chapter 4, including multivariate regression).
- *Multicollinearity* : If the predictor matrix is nearly singular, the predictors often become unstable (for example for the reduced rank regression methods).
- *Optimization* : If only few variables have to be considered, less experiments are needed in an optimization process. Additionally, each experiment can usually be done with smaller cost or amount of work.
- *Interpretation* : It is easier to interpret results for parsimonious models, that is for models with fewer variables.

It is sometimes argued that variable selection cannot be logically justified, unless a cost is explicitly associated with the measurement of variables (see, for example, Lindley, 1984). One might answer that the initially chosen predictors are already a subset of all plausible predictor variables (see, for example, Miller, 1984).

Although many strategies for selecting variables have been considered, such as forward selection, backward selection and best subset selection, only forward variable selection will be considered in the following, because of its computational ease and the production of nested models.

Forward variable selection is defined as follows :

First, that predictor variable is selected which gives the minimal estimated expected loss (see section 2.3.5) when used as a single predictor. Then, that variable is added which gives minimal expected loss when used as a predictor together with the previously selected variable. The variable selection ends, after all variables have been selected.

Forward variable selection orders the variables according to their importance. To obtain a prediction method, the number of selected variables has to be chosen, for example based on the minimal estimated expected loss (see section 2.3.5).

It is extremely time–consuming to use exploratory graphical methods (see section 2.4) for variable selection, especially if many variables have to be considered. However, once an ordering of the variables has been obtained, diagnostic methods can be used to select the appropriate number of variables.

Variable selection will be discussed in more detail in connection with the reduced rank regression models in chapter 4.

2.6. Assessment of the Goodness of Prediction of the Selected Model

If the formal criteria introduced in section 2.3 are used for model selection, then the finally selected prediction rule is the method with minimal estimated expected loss among a collection of candidates. If this minimal estimated expected loss is considered as an estimate of the goodness of prediction of the selected model, then its performance is usually overestimated (see, for example, Miller, 1984). However, if the interest centers mainly on the choice of the prediction method and not on a realistic estimate of the expected loss, this is of no consequence.

If a more reliable estimate of the expected loss is needed, then the whole selection process has to be taken into consideration as well. Since analytical estimates are very difficult to obtain, resampling methods such as crossvalidation (see section 2.3.5) appear to be the only alternative (see, for example, Roecker, 1991, and Breiman and Spector, 1992). For the heuristically motivated multivariate prediction methods in chapter 3, "double resampling" will be used, i.e. the dimensionality is selected by crossvalidation as well as the whole procedure of selecting the dimension (see Stone, 1974). For model based methods, the alternative is to select the model based on asymptotic theory (see section 2.3.5), and only to evaluate the selection procedure by resampling methods. For example in reduced rank regression (see chapter 4) the dimension is selected based on the asymptotic estimate of the information criterion IC or the mean squared error of prediction MSEP criterion, and the goodness of prediction of the selected model is evaluated by crossvalidation.

3.
Heuristic Multivariate Prediction Methods

3.1. Introduction

Multivariate prediction methods allow the prediction of a multivariate response for given multivariate predictor. A general overview of linear multivariate prediction methods has been given in chapter 2. Unless the predictors are set according to an experimental design, the predictors are correlated. The presence of severe multicollinearity among the predictor variables is the cause for instabilities of prediction methods such as multivariate regression and the reduced rank regression methods discussed in chapter 4. Variable selection procedures (see section 2.5) are one possible solution to this problem. Alternatively, the prediction methods can be modified, motivated by Bayesian analogies, to obtain more stable ridge regression like methods (see Lindley and Smith, 1970), a possibility that will be noted in chapter 5. Some further approaches to deal with multicollinearity are given, for example, in Toutenburg (1992, pp.46), among which are the performing of additional experiments, the use of prior information and the consideration of exact linear restrictions.

An other approach to deal with multicollinearity is to extract orthogonal "latent variables" from the predictor variables, which then can be used as new predictor variables in ordinary multivariate regression. Two such prediction methods are principal component regression (PCR) and partial least squares (PLS), which are described in detail in sections 3.2 and 3.3. These prediction methods are heuristic in the sense that it seems difficult to justify them as estimators of specific models (see Frank and Friedman, 1993). Both methods can be combined in a continuum of different prediction methods, called continuum regression (Stone and Brooks, 1990, and Sundberg, 1993), but will be discussed separately in the following.

In section 3.4 numerical prediction criteria (see section 2.3) are used for the selection of the number of latent variables, i.e. the dimensionality of the models. Finally, these multivariate prediction methods are illustrated by an example in section 3.5, using both numerical criteria (section 3.4) and graphical methods (section 2.4).

3.2. Principal Component Regression

Principal component regression (PCR) is well known in the statistical literature, going at least back to Massy (1965); see also Toutenburg (1992, pp.80). It consists of a regression of the first few principal components of the predictors on the responses.

Considering the setting discussed in section 2.2, suppose that X is an $n \times p$ known fixed centered matrix of predictors and Y is an $n \times q$ matrix of observed responses. Then the prediction rule (compare with (2.2.3)) defined by principal component regression for given dimension d is defined as follows.

(3.2.1) $\hat{y}' = \hat{\mu}(Y;X)' + x'\hat{M}(Y;X)$, where
$\hat{\mu}(Y;X) := 1/n \; Y'1_n$ is the mean of the response variables,
$\hat{M}(Y;X) := V_d S_d^{-1} U_d' \; (Y - 1_n\hat{\mu}(Y;X)')$ is the coefficient matrix and
V_d, S_d, U_d are obtained from a singular value decomposition of X.

The singular value decomposition of the matrix X is given by

(3.2.2) $X = USV'$, with
column orthogonal $n \times p$ matrix U,
orthogonal $p \times p$ matrix V and
diagonal $p \times p$ matrix S with decreasing positive diagonal elements.

V_d and U_d consist of the first d columns of V and U, respectively, and S_d consists of the first d rows and columns of S. The columns of U_d are called principal components.

Note that $X\hat{M}(Y;X) = U_d U_d{}' \ (Y - 1_n\hat{\mu}(Y;X)')$, i.e. (3.2.1) corresponds in fact to an ordinary multivariate regression of the first d principal components of the predictor matrix on the responses. The heuristic motivation for the use of principal component regression is that directions in the predictor space with small variation are considered as noninformative, and hence can be discarded. Since principal component analysis is not invariant under scaling of the predictors, different solutions are obtained depending on whether the predictors are standardized or not. If the predictor variables are measured in incomparable scales, then these variables are usually standardized to have a standard deviation of 1. If the variables are all measured in the same units, for example as absorptions for different wavelengths (spectral data), a standardization may not be necessary. An algorithm to calculate the coefficient matrix in (3.2.1) is given in the appendix.

3.3. Partial Least Squares

The partial least squares (PLS) methodology was developed by Wold (1982) as an algorithm for analyzing relationships between variables, especially in econometrics and the social sciences. However, the partial least squares algorithm considered in the chemometrics and statistical literature is a modification of this algorithm which is perhaps best regarded as a separate development (see, for example, Höskuldsson, 1988, Frank and Friedman, 1993, Helland, 1990,1992).

Using the same setting as in the previous section 3.2, the prediction rule defined by partial least squares for given dimension d is defined as follows.

(3.3.1) $\hat{y}' = \hat{\mu}(Y;X)' + x'\hat{M}(Y;X)$, where
$\hat{\mu}(Y;X) := 1/n \ Y'1_n$ is the mean of the response variables,
$\hat{M}(Y;X) := U \ R \ Z_d{}' \ (Y - 1_n\hat{\mu}(Y;X)')$ is the coefficient matrix and U, R, Z_d are obtained from a partial least squares analysis of X,Y.

The d–dimensional partial least squares analysis of X,Y is technically defined by the following sequential algorithm.

Define the centered matrices $\tilde{X} := X$ and $\tilde{Y} := (Y - 1/n \ 1_n 1_n{}'Y)$.

Then, for $i = 1, 2, \dots, d$, the following iterative procedure is applied

Let the p–dimensional vector U_i be the first principal component of $\tilde{X}'\tilde{Y}$.

Redefine the matrices $\tilde{X}$ and $\tilde{Y}$ as

$\tilde{X} := \tilde{X} - \tilde{U}_i\tilde{U}_i'\tilde{X}$ and $\tilde{Y} := \tilde{Y} - \tilde{U}_i\tilde{U}_i'\tilde{Y}$, where

$\tilde{U}_i := (\tilde{X}U_i)/\sqrt{(\tilde{X}U_i)'(\tilde{X}U_i)}$.

Define the column orthonormal $p \times d$ matrix $U := (U_1 \; \dots \; U_d)$ and the $n \times d$ matrix $Z_d := UR$, where R is an upper triagonal matrix chosen such that Z_d is column orthonormal. The columns of Z_d are called partial least squares components. Note that as for PCR, $X\hat{M}(Y;X) = Z_dZ_d' \; (Y - 1_n\hat{\mu}(Y;X)')$, i.e. (3.3.1) corresponds to an ordinary multivariate regression of the first d partial least squares components of the predictor matrix on the responses. A more detailed algorithm to calculate the coefficient matrix in (3.3.1) is given in the appendix.

Less technical, the partial least squares (PLS) methodology can be described as follows. The first partial least squares component is obtained in an analogous way as in canonical correlation analysis (CCA, see section 4.3.8.3), but maximizing covariance between $\tilde{X}$ and $\tilde{Y}$ instead of correlation : a linear combination of the predictors $\tilde{X}$ and of the responses $\tilde{Y}$ is searched for, such that the covariance between the two components is maximal. These linear combinations are the first components of a partial least squares analysis. To get the second components, the predictors $\tilde{X}$ and the responses $\tilde{Y}$ are replaced by residuals, which are obtained as follows. The responses $\tilde{Y}$ are replaced by the residuals of a multivariate regression of the first component of the predictors on $\tilde{Y}$, and the predictors $\tilde{X}$ are replaced by the residuals of a multivariate regression of the first component of the predictors on $\tilde{X}$. Then again linear combinations of $\tilde{X}$ and $\tilde{Y}$ searched for which maximize covariance, and so on.

Note that partial least squares is not invariant under scaling of both the predictors and the responses. As for principal component regression it depends on the problem whether a scaling is appropriate or not.

It is still an open question, for which statistical models the maximization of the covariance between predictors and responses may be useful (see, for example, Helland, 1990, 1992). However, extensive simulation studies (see Frank, 1991, Frank and Friedman, 1993) suggest that partial least squares and principal component regression have often a better prediction power than multivariate regression. This is supported as well by many applications to real data, especially in chemometrics, where the goodness of prediction is measured by the mean squared error of prediction, using crossvalidation (see section 2.3.5.1). Hence these methods seem to be at least useful for comparison with more model oriented methods.

3.4. Dimension Selection

Principal component regression (PCR) and partial least squares (PLS) define a prediction method once their dimensionality has been selected. Dimension selection is a special case of model selection discussed in section 2.5. Using the prediction criteria of section 2.3, the dimension minimizing the estimated expected loss is chosen. Since the multivariate prediction methods PCR and PLS are only heuristically motivated and no statistical model is associated with them, density prediction (see section 2.2) seems difficult to justify. Therefore only point prediction (see section 2.2) and the associated criterion, the mean squared error of prediction (MSEP, see section 2.3) will be considered in the following. For PLS it is difficult to derive asymptotic results for the estimation of the expected loss (see section 2.3.5.1). Hence, crossvalidation is usually suggested in the literature for the selection of dimensionality (see section 2.3.5.1 and, for example, Wold, 1977, Stone and Brooks, 1990).

For PCR an asymptotic estimate can be derived as follows. Considering the framework of section 2.3.5.1 with $Y = Y_0 + E$, one has

$$\hat{M}(Y) = PY_0 + PE\,,\ \text{where}\ P := V_d S_d^{-1} U_d'\,.$$

As for multivariate regression, the first order Taylor approximation is exact with $C(Y_0) = (I_q \otimes P)$. Hence

$$-\,1/2\ \delta = \mathrm{tr}[(\Sigma\Gamma^{-1} \otimes XV_d S_d^{-1} U_d')] = \mathrm{tr}(\Sigma\Gamma^{-1}) \cdot d\,.$$

An unbiased estimate of the error covariance matrix can be obtained if it is assumed that the full model with maximal dimension, i.e. the multivariate regression model is essentially correct. Then the covariance of the residuals is an estimate of Σ. More precisely

$\hat{\Sigma} := (Y_c - U_p U_p' Y_c)'(Y_c - U_p U_p' Y_c)/n$, where

$Y_c = (Y - 1/n \; 1_n 1_n' Y)$ is the centered response matrix.

The asymptotic estimate of the MSEP is now given by

(3.4.1) $MSEP_{ASY} := MSEP_{FIT} + 2\, d \cdot tr(\hat{\Sigma}\Gamma^{-1})$.

As discussed in section 2.6, these estimates of the MSEP might still be too optimistic, since the "selection bias" (Miller, 1984) is not taken into account. Realistic estimators of the predictive power of the selected method are only obtained if the selection of the dimension is automatized and hence part of the prediction method. For PCR either the asymptotic or the crossvalidation estimate of the MSEP can be used for automatic dimension selection, while for PLS only the selection by the crossvalidation estimate seems possible. To evaluate the complicated prediction method with automatic dimension selection, only resampling methods are feasible.

In addition to these formal procedures exploratory graphical methods can be used for dimension selection (see section 2.4).

3.5. Example

For the insecticide data from the QSAR area described in section 1.6, the mean squared error of prediction (MSEP, (2.3.7)) is estimated for both methods and all possible dimensions. The weight matrix defined in (2.3.2) is used, so that the MSEP is equivalent to the mean of the individual R^2 values (R^2_{mean}). For principal component regression (PCR), the fit, asymptotic and crossvalidation estimates and for partial least squares (PLS), the fit and crossvalidation estimates are calculated according to section 2.3.5.1 and 3.4. They are shown in Table 3.5.1.

One sees that for PCR, at least 6 dimensions seem to be necessary for a good prediction. According to both the asymptotic and the crossvalidation estimate, 8 dimensions are considered to be optimal. For PLS, at least 2 dimensions are needed and 5 dimensions seem to be optimal. Comparing the MSEP estimates obtained by crossvalidation, one sees that both methods seem to have a very similar predictive power, which however is much smaller than suggested by the fit estimate.

Table 3.5.1 Estimated prediction criteria mean R^2 (maximal values in bold face)

Principal component regression (PCR)									
MSEP	DIM1	DIM2	DIM3	DIM4	DIM5	DIM6	DIM7	DIM8	DIM9
fit	0.029	0.029	0.068	0.276	0.546	0.686	0.716	0.763	0.767
asy	0.003	-0.023	-0.010	0.172	0.416	0.531	0.535	**0.556**	0.534
cv	-0.141	-0.248	-0.289	-0.253	0.123	0.405	0.396	**0.417**	0.368

Partial least squares (PLS)									
MSEP	DIM1	DIM2	DIM3	DIM4	DIM5	DIM6	DIM7	DIM8	DIM9
fit	0.580	0.675	0.701	0.729	0.751	0.758	0.761	0.763	0.767
cv	0.237	0.372	0.373	0.412	**0.436**	0.433	0.414	0.410	0.368

To obtain more realistic estimates of the performance of the selected method, methods with automated dimension selection have to be considered as discussed in section 3.4. In Table 3.5.2 one sees that for PCR, the selected model seems to have in fact a smaller predictive power than one would have naively expected from looking only at Table 3.5.1.

Table 3.5.2 Estimated prediction criteria (mean R^2)

PCR with automatic dimension selection							
by MSEPasy :	MSEPcv 0.385						
	selected dimensions	6	7	8			
	number of selections	1	1	25			
by MSEPcv :	MSEPcv 0.310						
	selected dimensions	6	7	8			
	number of selections	8	1	18			

PLS with automatic dimension selection							
by MSEPcv :	MSEPcv 0.438						
	selected dimensions	3	4	5	6	7	8
	number of selections	2	1	11	10	2	1

Additionally, the dimensions selected by the automated dimension selection method for each of the left–out observations are summarized in Table 3.5.2. As expected from Table 3.5.1, between 6 and 8 dimensions have been selected for PCR, and most often the model with 8 dimensions. Automatic dimension selection by the asymptotic criterion seems to give a better prediction method than selection by the crossvalidation criterion. In contrast to PCR, for PLS almost no differences in the predictive power of the automated procedure and the selected method are observed in Table 3.5.2. However, the selected number of dimensions varies more than for PCR, ranging between 3 and 8 dimensions.

In Table 3.5.3 the coefficient matrices for the two selected methods are shown. The difference between these coefficient matrices is small.

Table 3.5.3 Coefficient matrices M of the prediction models

	PCR 8 dim.		PLS 5 dim.	
	ADULTS	LARVAE	ADULTS	LARVAE
CONST	-1.6845	-1.7508	-1.6845	-1.7508
I1	-0.2800	-0.2167	-0.2852	-0.2245
I2	0.2938	0.3442	0.2877	0.3272
IP23	-0.0239	0.1436	-0.0183	0.1276
Pi	0.0970	0.0886	0.0839	0.0717
F	-0.0319	-0.0996	-0.0727	-0.1791
R	-0.1953	-0.2114	-0.1922	-0.2015
MR	0.1390	0.3653	0.0781	0.2357
L	-0.0426	-0.1207	-0.0099	-0.0795
B1	-0.0780	-0.2631	-0.0461	-0.1812

That the two methods are very similar can be seen as well in Figures 3.5.1 and 3.5.2, where the predictors are compared graphically. Figure 3.5.1 is a parallel display of the two prediction methods, where cross-validation predictions of the responses are plotted versus the observed responses. In an interactive graphical environment, the plots are linked, that is the points in the different scatterplots corresponding to the same observation can be easily identified. This helps to see, whether an observed response which is badly predicted by one prediction method is badly predicted by the other prediction method as well. Figure 3.5.2.

shows a static interpolation of the crossvalidation residuals obtained by the two prediction methods. For each method, the crossvalidation residuals are calculated as the differences between observed responses and crossvalidation predictions of the responses and displayed as a scatterplot. The two scatterplots are then superimposed and the corresponding observations are joined by line segments. When the two prediction methods are similar, the line segments will all be short. In a dynamic graphical environment, the two scatterplots corresponding to the two methods are not superimposed (static interpolation) but dynamically interpolated, i.e. a series of scatterplots is shown, starting with the scatterplot of residuals of one prediction model and ending with the scatterplot of residuals of the other prediction model.

Figure 3.5.1 *Crossvalidation predictions for the two selected methods PCR (8 dim.) and PLS (5 dim.) : scatterplots of predicted and observed values. Ideally, the points lie on the diagonal.*

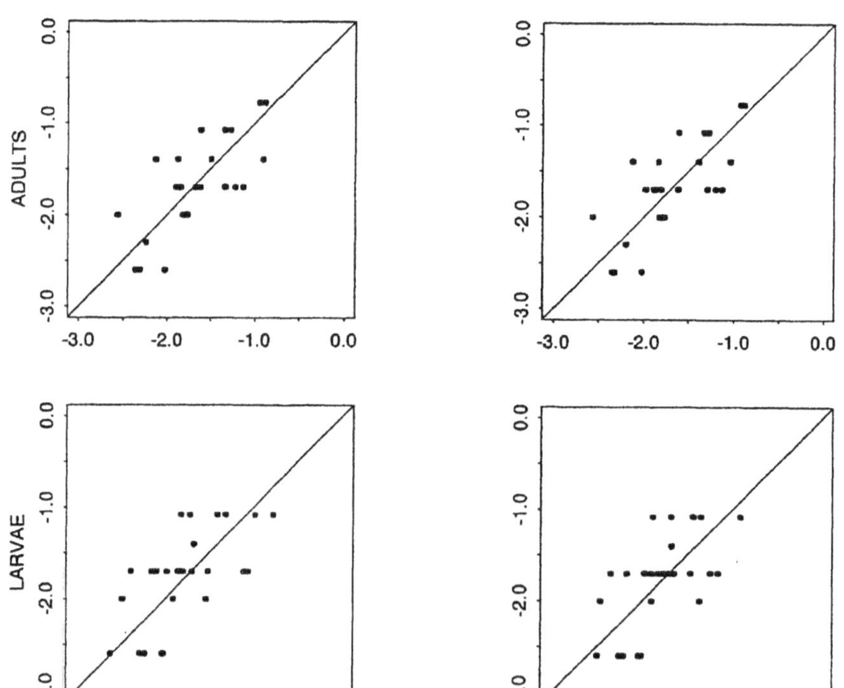

Figure 3.5.2 *Crossvalidation residuals for the two selected methods PCR (8 dim.) and PLS (5 dim.) : static interpolation of corresponding observations (PCR :• ; PLS : .). For two very similar methods, the line segments are small.*

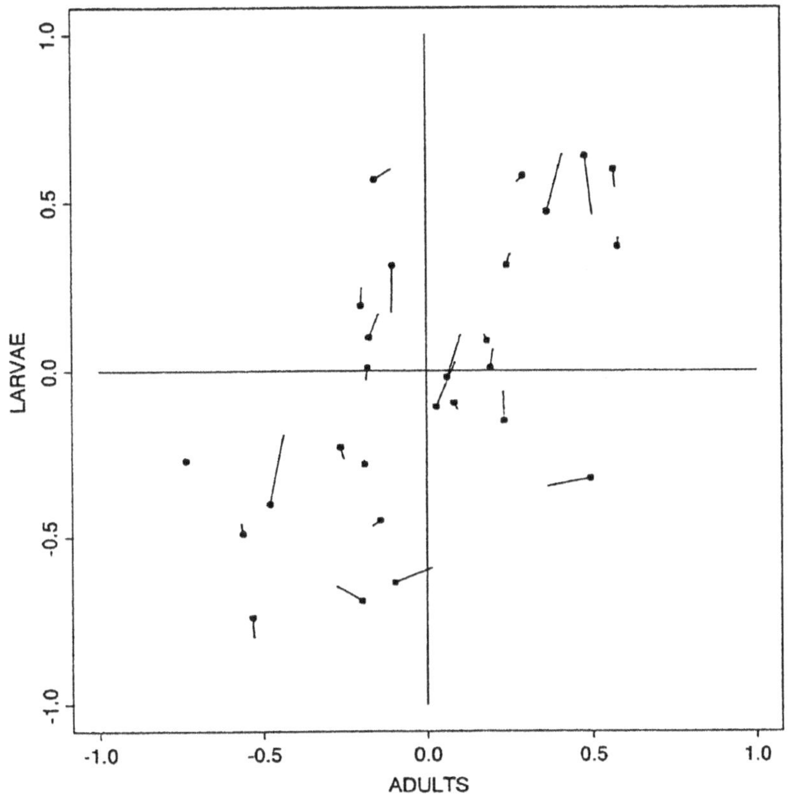

In section 1.6, it was noted that some test substances are available as well, which have not been used for the building the multivariate prediction models. Comparing the predicted and observed values for this test dataset (see Table 1.6.1) in Table 3.5.4, one sees that the predictions are acceptable with the exception of two substances, namely compound IIia and IIIia . As some additional experiments showed, these substances are unstable under UV–light in contrast to all other substances. This explains their bad predictability. For a more detailed description of the chemodynamic behaviour of these substances and the experimental setup, see again Pascual et al. (1994).

Table 3.5.4 Predictions for the test data (outlier substances in bold face)

	OBSERVED		PCR 8 dim. PREDICTED		PLS 5 dim. PREDICTED	
	ADULTS	LARVAE	ADULTS	LARVAE	ADULTS	LARVAE
IIca	-2.0	>-1.7	-1.656	-1.955	-1.686	-1.977
IIcb	>-1.7	-2.0	-2.225	-2.395	-2.266	-2.433
IIia	**-2.6**	**-2.6**	**-0.781**	**-0.591**	**-0.826**	**-0.702**
IIIia	**>-1.7**	**-2.0**	**-0.204**	**0.084**	**-0.261**	**-0.059**
IIIka	>-1.7	-1.7	-1.051	-1.100	-1.056	-1.137
IIIla	>-1.7	>-1.7	-1.099	-1.108	-1.058	-1.002
IIIma	>-1.7	>-1.7	-1.140	-1.019	-1.259	-1.270

4. Classical Analysis of Reduced Rank Regression

4.1. Introduction

In contrast to the heuristic multivariate prediction methods discussed in chapter 3, the reduced rank regression methods are a modeling approach to multivariate prediction. The methods are closely linked to many standard linear multivariate methods such as canonical correlation analysis (Tso, 1981), redundancy analysis (van den Wollenberg, 1977), canonical discriminant analysis (Izenman, 1986) and principal component analysis (Izenman, 1975). Slightly less close relationships can be established to other methods such as factor analysis (Jöreskog, 1967) and LISREL (vanderLeeden, 1990). Additionally, the reduced rank regression approach is itself very natural in QSAR analyses, as shown in the next section 4.2. Applications to other areas in biometrics (Ryan et al., 1992), chemometrics (Davies and Tso, 1982), econometrics (Anderson, 1951, Jöreskog and Goldberger, 1975, Gudmundsson, 1977, Geweke, 1993) and other sciences (Velu, 1991, Glasbey, 1992) can be justified as well.

Reduced rank regression models are formally defined in section 4.3, where maximum likelihood estimates for these models and their asymptotic covariances are derived. After a discussion of possible extensions in section 4.4, the prediction criteria presented in chapter 2 are used for rank selection in section 4.5. Their ability to select the correct rank is investigated in a small simulation study. In the final section

4.6, the selection of predictor variables by prediction criteria is discussed and again some insights are gained by a simulation study. All methods are illustrated by applying them to the insecticide example from the area of quantitative structure–activity relationships presented in section 1.6.

4.2. QSAR : Biological Responses

Reduced rank regression models arise very naturally in a QSAR context. The discussion in chapter 1 is now taken up in more detail and it is argued that the proposed methods may be adequate to model quantitatively the relationship between (chemical) structure and (biological) activity.

As noted in chapter 1, the biological activity of a chemical substance is generally observed through a number of variables. There can be many reasons for not just using a single variable for the description of the biological activity of a chemical substance. In some cases a number of biological effects for which one would not a–priori expect any correlation may be of interest. As an example, the toxic side–effect of an anti–inflammatory drug is not thought to be related to its potential to inhibit inflammation. In other cases, the biologist or physicist has difficulties to decide which of the observable variables represents best biological activity, and instead of chosing one specific variable, selects a group of similar variables. For example, the damage of a weed can be characterized by the reduction in height or the size reduction of the leafs. Additionally, not only one species of weed, but a large number of different weeds can be used to measure biological activity. In the first case of unrelated responses, a two stage procedure seems adequate, which consists of first analyzing the responses individually, and then combining them by an additional criterion. For example, toxicity and inhibition of inflammation might first be modeled separately, and then merged by a desirability index, balancing toxic side–effects and beneficial aspects of the drug. But even in this case, a multivariate analysis can be adequate, if for example the different variables are measured in a common experiment, i.e. when the errors are not independent. In the second case, where the measured response variables are just indicators of the same biological effect, only a multivariate analysis is adequate, even if the errors can be assumed to be independent.

To be more specific, the biological effect of a substance is often due to the blocking of a specific enzyme in the biological system which in turn is for example essential for its growth. While the primary cause for the growth inhibition, i.e. the enzyme inhibition, is not directly observed, its effect is indirectly made visible by the measurement of the height or the leaf size of the plant. To be more precise, suppose that the unobserved latent variable z reflects the enzyme blocking capacity of a substance. This primary cause of biological activity is then observed through a number of response variables $y_1, y_2, ..., y_q$, which are, apart from an additive measurement error, functions of the (one–dimensional) latent variable z, so that formally, one can write

(4.2.1) $y_i = f_i(z) + e_i, \quad i=1,...,q$, where
e_i is a random variable with zero expectation (error),
z is a fixed latent variable and
$f_i : R \rightarrow R$ is a function.

In vector form, one has

(4.2.2) $y = f(z) + e$, where
$y = (y_1, ..., y_q)$ is a q–dimensional random variable of responses,
$e = (e_1, ..., e_q)$ is a q–dimensional random variable of errors,
z is a fixed latent variable and
$f : R \rightarrow R^q$ is a function.

This relationship between observed responses and latent variable can be linearized to give

(4.2.3) $y' = \mu' + z\beta' + e'$, where
μ is the q–dimensional mean and
β is a q–dimensional vector of coefficients.

Note that this model is essentially a (one–dimensional) factor analysis model for the response variables (see Jöreskog, 1967). The parameters of this model may be interpreted as follows : due to different units, different strengths of reaction, etc., the responses are scaled and shifted versions of the latent variable.

Clearly, the physico–chemical properties of a substance are responsible for its enzyme blocking capacity. Hence, if all the relevant physico–chemical properties are given by the predictor variables $x_1, ..., x_p$, the latent variable z is a function of these properties.

Formally one can write,

(4.2.4) $z = g(x)$, where
$x = (x_1, ..., x_p)$ is a p–dimensional fixed vector of predictors and
$g : R^p \rightarrow R$ is a function.

A linearized version of this relationship is then given by

(4.2.5) $z = x'\alpha$, where
α is a p–dimensional vector of coefficients.

Putting the two linearized parts (4.2.3) and (4.2.5) together, a plausible model for the relationship between observed responses and predictors is given by

(4.2.6) $y' = \mu' + x'\alpha\beta' + e'$.

If one combines the two coefficient vectors α, β into a $p \times q$ coefficient matrix $M := \alpha\beta'$, one sees that this model is a special case of a standard multivariate regression model, namely one with the additional restriction that the coefficient matrix M is of reduced rank 1 instead of full rank. Hence one can write model (4.2.6) as

(4.2.7) $y' = \mu' + x'M + e'$,
$\text{rank}(M) = 1$.

Multivariate regression models with a rank restriction on the coefficient matrix are called reduced rank regression models. For another motivation for the use of latent factors in a regression context, see for example Jöreskog and Goldberger (1975).

4.3. Reduced Rank Regression Models

Reduced rank regression models have been introduced by Anderson (1951) while the now generally used name is due to Izenman (1975). A detailed history of the development of these models can be found in vanderLeeden (1990). Some additional references are given in Velu (1991). In the following, first the basic model is presented in section 4.3.1, emphasizing the importance of an adequate parametrization (section 4.3.2). Then the advantages of the model in comparison with multivariate regression are discussed (section 4.3.3) as well as an important geometrical aspect (section 4.3.4). In sections 4.3.5 to 4.3.8, maximum likelihood estimates for three important reduced rank regression models are given. In the following section 4.3.9, the asymptotic covariance matrix of the estimates is derived, which can be used to calculate approximate prediction regions. Finally, an example illustrates the methods in section 4.3.10.

4.3.1. Model

Multivariate regression is a generalization of univariate regression to more than one response variable. The basic model is given by

(4.3.1) $y' = \mu' + x'M + e'$, where
M is a $p \times q$ matrix of unknown parameters (coefficient matrix),
μ is a q–dimensional vector of unknown parameters (mean),
x is a p–dimensional known fixed vector (predictor),
y is a q–dimensional random variable (response) and
e is a q–dimensional random variable with zero expectation (error).

Reduced rank regression models differ from multivariate regression by additional rank constraints on the coefficient matrix M. The standard reduced rank regression model can be written as

(4.3.2) $y' = \mu' + x'M + e'$, where
M is a $p \times q$ matrix of coefficients of fixed rank $r < \min(p, q)$.

Suppose that for i=1, ... , n , the independent q–dimensional random response variables y_i are defined by the reduced rank regression model (4.3.2) with corresponding p–dimensional known fixed predictor variable x_i and i.i.d. q–dimensional random error variables e_i. The reduced rank regression model can then be written in matrix form as

(4.3.3a) $Y = 1_n\mu' + XM + E$, where
M is a $p \times q$ matrix of coefficients of fixed rank $r < \min(p, q)$,
$X = (x_1' \dots x_n')'$ is an $n \times p$ known fixed matrix of predictors,
$Y = (y_1' \dots y_n')'$ is an $n \times q$ random matrix of responses and
$E = (e_1' \dots e_n')'$ is an $n \times q$ random matrix of errors.

It is assumed in the following that the matrix X has full rank and, without loss of generality, that the predictor matrix X is centered, i.e. that $X'1_n = 0_p$. It is convenient to further assume that the errors are normally distributed, i.e. that

(4.3.3b) $E \sim N(0_{n \times q}, \Sigma \otimes I_n)$ matrix normal distribution, where Σ is a positive definite $q \times q$ matrix of unknown parameters (error covariance matrix).

An $n \times q$ random matrix U has a matrix normal distribution denoted by $N(M, \Omega)$, if its vectorization $vec(U)$ has a multivariate normal distribution with nq–dimensional mean vector $vec(M)$ and $nq \times nq$ covariance matrix Ω, where M is an $n \times q$ matrix.

4.3.2. Parametrization

Up to now, the reduced rank regression model has been defined as a multivariate regression model with rank restrictions on the coefficient matrix. However, for estimation purposes, an explicit parametrization of the coefficient matrix of reduced rank is necessary. Obviously, every $p \times q$ matrix M of rank r can be written as the product $M = AB'$ of a $p \times r$ matrix A and a $q \times r$ matrix B, where both A and B have rank r. Unfortunately, this decomposition is not unique, since for every invertable $r \times r$ matrix F, one has

$$M = AB' = AFF^{-1}B' = (AF)(BF'^{-1})'.$$

Essentially three different possibilities have been suggested in the literature to resolve this ambiguity. A similar but more general discussion will be needed for the Bayesian analysis of reduced rank regression models in chapter 5.

a) Parametrization I

(4.3.4) $M = AB'$, $A = (I_r \ \ \tilde{A}')'$, B arbitrary, where $\tilde{A}$ is an arbitrary $(p - r) \times r$ matrix.

While this obviously solves all uniqueness problems, not every matrix M of rank r can be written as the product of a matrix A of the form (4.3.4) and an arbitrary matrix B. To see this, suppose that the first r rows of the matrix M are zero. This is only possible if B is identical to a zero matrix, hence that all elements of M are zero. Of course, the set of matrices M of rank r that cannot be decomposed as (4.3.4) is a set of measure 0. Nevertheless in practice, matrices near to this set may occur, posing serious numerical problems which can only be remedied by awkward repeated reordering of the predictor variables (see the suggestions in Ahn and Reinsel, 1988, Geweke, 1993). However, this parametrization I (4.3.4) has some advantages in deriving asymptotic results (see section 4.3.10 and Ryan, 1992).

b) Parametrization II

(4.3.5) $M = AB'$, $B = (I_r \ \tilde{B}')'$, A arbitrary, where
$\tilde{B}$ is an arbitrary $(q - r) \times r$ matrix.

The same numerical problems already discussed above are encountered. The only difference is that in applications, the response variables have to be repeatedly reordered instead of the predictor variables.

c) Parametrization III

(4.3.6) $M = AB'$, $A'X'XA = I_r$, B arbitrary.

This parametrization III of the parameters has numerous advantages over the parametrizations I and II given above. First, a reordering of the response or predictor variables will not change the outcomes. Second, the reduced rank regression model has been motivated in section 4.2 by assuming that the expected value of the responses Y is linearly dependent on latent factors Z, which in turn are linear combinations of the predictor variables X, i.e. $E(Y) = ZB$, $Z = XA$. If the parametrization III (4.3.6) is chosen, this corresponds to the requirement that the latent factors Z are orthonormal, a requirement familiar to factor analysis. The orthonormality of the latent factors makes their interpretation much easier, since no correlations among them are present. Additionally the interpretation of the coefficient B is straightforward, since for fixed A, the reduced rank regression model then corresponds to a multivariate regression model with orthonormal predictors.

Another advantage of parametrization III is apparent if graphical methods are used. Since the latent variables are orthonormal, all linear structure is removed in the graphical display and nonlinearities are easier to detect.

Finally, as seen later, reduced rank regression models are linked to many multivariate analysis methods such as canonical correlation analysis (section 4.3.8.3), redundancy analysis (section 4.3.8.2), etc., where usually the same parametrization III (4.3.6) is used and the latent variables are called scores. A disadvantage of this parametrization III is the non–linearity of the parameter space making it somewhat difficult to calculate and interpret confidence regions.

There is still a uniqueness problem with the parametrization III (4.3.6), typical for most multivariate methods. Note that for every orthogonal $r \times r$ matrix O, one has

$$M = (AO)(BO)' = AOO'B' = AB' \text{ and}$$
$$(XAO)'(XAO) = O'(XA)'(XA)O = O'O = I_r,$$

i.e. a simultaneous orthogonal rotation of both parameters A, B is possible without changing the coefficient matrix M and violating the constraints. The rotational indeterminacy is perhaps most intensely studied in a factor analysis context, where various rotation methods are proposed, with the goal of obtaining simple coefficient matrices (see, for example, Kaiser, 1958). Since this rotational indeterminacy poses no problems in the estimation of the parameters, this non–uniqueness is not explicitly resolved in the literature.

If an explicit restriction is needed, for example in a resampling context, or if similar datasets have to be compared, the following additional requirement solves this problem. The parameter matrix A has to be such that

(*) $\| (A - A_0)(X'X)^{1/2} \| \leq \| (A\,O - A_0)(X'X)^{1/2} \|$
for all orthogonal matrices O, where A_0 is some fixed matrix with $A_0{}'X'XA_0 = I_r$.

In a resampling context, A_0 could be chosen to be a maximum likelihood estimate from the original dataset. If several similar datasets are compared, then A_0 could be chosen as a maximum likelihood estimate obtained from one of the datasets.

The actual rotation can be calculated analytically by a (generalized) Procrustes analysis (see Sibson, 1978) as follows. Suppose that $\tilde{A}$ is a maximum likelihood estimate not fulfilling the requirement (*). Let $\tilde{A}'X'XA_0 = U\,S\,V'$ be the singular value decomposition of $\tilde{A}'X'XA_0$, with column orthogonal matrix U, diagonal square matrix S and orthogonal matrix V. Then $A = \tilde{A}V$ is a maximum likelihood estimate as well and fulfills the requirement (*). The name Procrustes stems from the fiend in Greek mythology, who stretched or compressed his guests to fit exactly in his bed.

4.3.3. Reduced Rank Regression or Multivariate Regression ?

One might perhaps argue that since the reduced rank regression model is just a multivariate regression model with rank constraints on the coefficient matrix, ignoring these restrictions is still acceptable. There are three main reasons why this is not appropriate, if the rank constraints are justified :

– *interpretation :* The number of parameters of the coefficient matrix M is drastically reduced if a reduced rank regression model is used instead of a multivariate regression model. For example, if there are p=10 predictor variables and q=5 response variables following a reduced rank regression model of rank r=1, then the number of parameters goes drastically down from pq=50 to $(pr + qr - r^2)$=14. Hence if the reduced rank regression model is essentially correct, many redundant parameters do not have to be considered.

– *optimization* : In a predictive setting, where a better combination of the predictor variables with respect to all response variables is looked for, a multivariate regression model could give conflicting optimization strategies. If these differences are due to estimation errors and a reduced rank regression model of rank r=1 is adequate, such spurious conflicts can be avoided.

– *precision* : Since a smaller number of parameters has to be estimated for the reduced rank regression models, one might expect that the estimates are more precise. In fact this can be proven for large samples (see section 4.3.10) and is plausible for small samples.

4.3.4. The Geometry of Reduced Rank Regression Models

A geometric interpretation of the reduced rank regression model helps to understand graphical displays of the prediction models. Detailed discussions can be found, for example, in Tso (1981), Davies and Tso (1982), terBraak (1990) and vanderLeeden (1990). The most important geometrical aspect of reduced rank regression models is the following : if an r–dimensional reduced rank regression method is used for prediction, the predicted values lie all in an r–dimensional subspace of the responses. This is now illustrated with the data form example 1.6. A 1–dimensional reduced rank regression model with unstructured error covariance matrix (see following sections) is estimated and both the predicted and observed values are displayed as a scatterplot in Figure 4.3.1. Observed and predicted values are joined by a line segment to be able to identify which are the corresponding predictions. Obviously, all predicted values lie on a line, i.e. on a 1–dimensional subspace of the responses. Note as well, that the line segments joining observed and predicted responses are generally not perpendicular to the 1–dimensional subspace.

Figure 4.3.1 *One dimensional reduced rank regression model : predicted (•) and observed (●) values are joined. The predicted values lie on a line.*

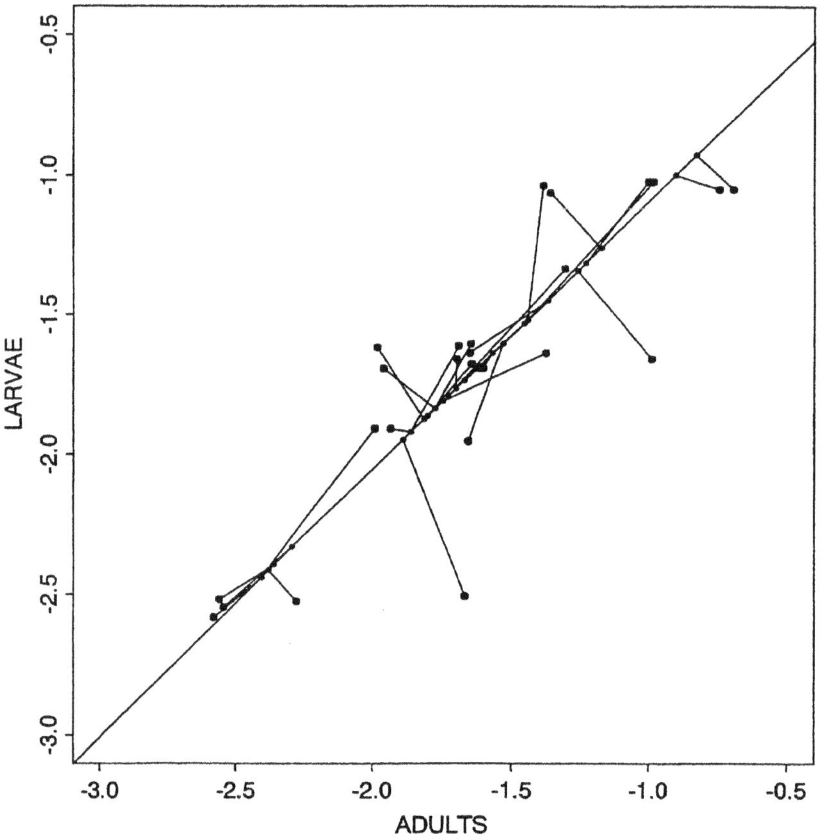

4.3.5. Likelihood

Model (4.3.3a), (4.3.3b) can be written as

(4.3.7) $Y \sim N(1_n\mu' + XAB', \Sigma \otimes I_n)$, where
A is a $p \times r$ matrix of unknown parameters of rank r with $(XA)'(XA) = I_r$,
B is a $q \times r$ matrix of unknown parameters of rank r.

The average log–likelihood of the observed data Y is then given by

$$
\begin{aligned}
(4.3.8)\quad & l(Y; A, B, \Sigma, \mu) \\
& = -q/2 \log(2\pi) - 1/2 \log|\Sigma| - 1/2 \mathrm{tr}[S(Y; A, B, \mu)\Sigma^{-1}], \\
& \text{where } S(Y; A, B, \mu) \\
& \qquad = (Y - 1_n\mu' - XAB')'(Y - 1_n\mu' - XAB')/n.
\end{aligned}
$$

Hence, the maximum likelihood estimates of the parameters A, B, Σ, μ are obtained as the solutions of the following optimization problem

$$
\begin{aligned}
(4.3.9)\quad & \log|\Sigma| + \mathrm{tr}[S(Y; A, B, \mu)\Sigma^{-1}] \rightarrow \min, \\
& \text{with the constraint } (AX)'(AX) = I_r .
\end{aligned}
$$

4.3.6. Error Structure

Up to now, no specific structure has been put on the error covariance matrix Σ in (4.3.7). However in most QSAR studies (see chapter 1) more detailed information about the error structure is available. In the following, four different error structures will be discussed, which are both easy to identify in a specific application, and cover most encountered situations.

a) Σ known :
If the responses are measured in a routine experiment, estimates of the size of the measurement error are sometimes available from past replicate experiments. If these estimates are precise enough, the error covariance matrix Σ can be treated as a fixed matrix. An alternative would be to use Bayesian methods to incorporate the variability of this error estimate (see chapter 5).

b) Σ proportional to the identity matrix :
If all response variables are measured by the same measurement procedure, it is often plausible to assume that all measurement errors are of the same size. If additionally the measurements for each response are taken in separate experiments, the errors between responses are independent, so that the error covariance matrix can be assumed to be proportional to the identity matrix.

c) Σ unknown positive definite :
If the response variables are measured in a common experiment, the errors are usually correlated. In this case, more detailed information about the error structure is seldom available so that it has to be assumed that the error covariance matrix is an arbitrary positive definite matrix.

d) Σ diagonal :
If the responses are measured in different experiments, but due to different measurement procedures the measurement errors are not of the same size, it seems plausible to assume that the error covariance matrix is diagonal.

Different procedures to find the maximum likelihood estimates of the parameters A, B, Σ, μ in model (4.3.7) have to be used, depending on which of the four error models is assumed (see section 4.3.8.

4.3.7. Maximum Likelihood Estimation of the Parameters B, Σ, μ

For given parameter A, the reduced rank regression model (4.3.7) is a standard multivariate regression model with predictor matrix $Z := XA$. Hence, the maximum likelihood estimates for the parameters B, Σ, μ are easily obtained, once the maximum likelihood estimate $\hat{A}$ of A is known. The maximum likelihood solutions are given by the parameters solving the optimization problem (4.3.9). Since the predictor matrix X is centered, i.e. $X'1_n = 0$, it is easy to see that the maximum likelihood estimate of μ is given by

$$(4.3.10) \quad \hat{\mu} := 1/n \ \ Y'1_n \ .$$

The maximum likelihood estimate of the error covariance matrix depends on the assumed error covariance structure. Again it is easy to see, that once the maximum likelihood estimates of the parameters A, B, μ are given by $\hat{A}, \hat{B}, \hat{\mu}$, the maximum likelihood estimate of Σ is

$$(4.3.11) \quad \text{a)}\ \Sigma = \Sigma_0 \text{ known}: \quad \hat{\Sigma} := \Sigma_0\,;$$

$$\text{b)}\ \Sigma = \sigma^2 I_q: \quad \hat{\Sigma} := s^2 I_q\ , \text{ where}$$

$$s^2 := 1/q \sum_{i=1}^{q} (S(Y; \hat{A}, \hat{B}, \hat{\mu}))_{ii}\ ;$$

$$\text{c)}\ \Sigma \text{ arbitrary}: \quad \hat{\Sigma} := S(Y; \hat{A}, \hat{B}, \hat{\mu})\ ;$$

$$\text{d)}\ \Sigma = \mathrm{Diag}(\sigma_1^2, \ldots, \sigma_q^2):$$

$$\hat{\Sigma} := \mathrm{Diag}(s_1^2, \ldots, s_q^2)\ , \text{ where}$$

$$s_i^2 := (S(Y; \hat{A}, \hat{B}, \hat{\mu}))_{ii}\ .$$

The maximum likelihood estimate of the parameter B for given estimate $\hat{A}$ of A does not depend on the error covariance structure. From multivariate regression, it is known that the maximum likelihood estimate of B is given by

$$(4.3.12)\quad \hat{B} := [(X\hat{A})'(X\hat{A})]^{-1}(X\hat{A})'Y = \hat{A}'X'Y.$$

Note again the advantage of using the parametrization III (4.3.6), namely $(XA)'(XA) = I_r$.

The problem of finding the maximum likelihood estimates of the parameters in the reduced rank regression model (4.3.7) is now reduced to the problem of finding the maximum likelihood estimate of the parameter A . Different estimation procedures have to be used, depending on which of the four error structures in section 4.3.6 is assumed. They will be discussed in the following section 4.3.8.

4.3.8. Maximum Likelihood Estimation of the Parameter A

4.3.8.1. Known Error Covariance Matrix

The case of a fixed error covariance matrix Σ_0 has been considered among others by Izenman (1975), Davies and Tso (1982), van der Leeden (1990) and Schmidli (1992).

Using the conditional maximum likelihood estimates given by (4.3.10–12), the optimization problem (4.3.9) can be written as

(4.3.13) $\mathrm{tr}[(Y_c - XAA'X'Y_c)\Sigma_0^{-1}(Y_c - XAA'X'Y_c)'] \rightarrow \min$,
with the constraint $(XA)'(XA) = I_r$,
where $Y_c := Y - 1/n\ 1_n 1_n'Y$ is the centered response matrix.

Obviously, this corresponds to a weighted least squares problem given by

(4.3.14) $\| (Y_c - XAA'X'Y_c)\Sigma_0^{-1/2} \|^2 \rightarrow \min$, with the constraints $(XA)'(XA) = I_r$.

An analytical solution of (4.3.14) is, for example, given by van der Leeden (1990).

4.3.8.2. Error Covariance Matrix Proportional to the Identity Matrix

The case of an error covariance matrix proportional to the identity matrix has been considered by Davies and Tso (1982), van der Leeden (1990) and Schmidli (1992).

Using again the conditional maximum likelihood estimates in section 4.3.7, the maximum likelihood estimate of A is the solution of

$$(4.3.15)\quad \mathrm{tr}[(Y_c - XAA'X'Y_c)(Y_c - XAA'X'Y_c)'] \rightarrow \min, \\ \text{with the constraint}\ (XA)'(XA) = I_r .$$

Written as a least squares problem, one has

$$(4.3.16)\quad \| Y_c - XAA'X'Y_c \|^2 \rightarrow \min, \text{ with the constraint} \\ (XA)'(XA) = I_r .$$

This problem is a special case of (4.3.14) with $\Sigma_0 = I_q$. An analytical solution is given in the appendix.

It is worth noting, that redundancy analysis (RDA, van den Wollenberg, 1977), a multivariate method well known in the psychometric literature is equivalent to this special reduced rank regression model as noted by Israëls (1984). Originally, redundancy analysis was proposed as an asymmetric alternative to canonical correlation analysis (see section 4.3.8.3) : linear combinations of the predictor variables and linear combinations of the response variables are searched such that the redundancy is maximal, where the redundancy is the mean variance of the linear combinations of the response variables explained by the linear combinations of the predictor variables.

Formally, the linear combinations A, C are solutions to the problem

$$(4.3.17)\quad \mathrm{tr}(XA)'(Y_cC) \rightarrow \max, \quad \text{with the constraints} \\ (XA)'(XA) = I_r,\ C'C = I_r .$$

Redundancy analysis is called asymmetric, since different solutions are obtained when X and Y are exchanged, in contrast to canonical correlation analysis.

4.3.8.3. Unstructured Error Covariance Matrix

Reduced rank regression with unstructured error covariance matrix is the situation originally discussed by Anderson (1951) who derived the maximum likelihood estimates. Tso (1981) showed that canonical correlation analysis is equivalent to this special reduced rank regression model. Further discussions can be found in Izenman (1975), Velu et al. (1986), van der Leeden (1990) and Schmidli (1992).

Using again the conditional maximum likelihood estimates of section 4.3.7, the following problem has to be solved

$$(4.3.18)\quad |(Y_c - XAA'X'Y_c)(Y_c - XAA'X'Y_c)'| \rightarrow \min, \quad \text{with the constraint } (XA)'(XA) = I_r .$$

An analytical solution of this problem is given in the appendix.

As already noted, canonical correlation analysis (CCA) is equivalent to this reduced rank regression model. Canonical correlation analysis searches for linear combinations of the predictor variables and linear combinations of the response variables such that the correlation between them is maximal. Formally, the linear combinations A, C are solutions to the problem

$$(4.3.19)\quad \mathrm{tr}(XA)'(YC) \rightarrow \max, \quad \text{with the constraints } (XA)'(XA) = I_r ,\ (YC)'(YC) = I_r .$$

Note the difference to redundancy analysis, where the coefficients C are chosen to be orthogonal instead of the linear combinations YC.

As noted by Izenman (1975) and Velu et al. (1986), it is possible to formulate an equivalent weighted least squares problem, namely

$$(4.3.20)\quad \| (Y_c - XAA'X'Y_c)\Gamma^{-1/2} \|^2 \rightarrow \min, \quad \text{with the constraint } (XA)'(XA) = I_r , \quad \text{where } \Gamma = Y_c'Y_c \text{ or } \Gamma = (Y_c'Y_c - Y_c'X(X'X)^{-1}X'Y_c).$$

4.3.8.4. Diagonal Error Covariance Matrix

The case of a diagonal error covariance matrix is very natural for many applications. Nevertheless it seems that this situation is only considered in van der Leeden (1990) and Schmidli (1992). One reason might be that no analytical solution of the maximum likelihood estimation problem seems to exist. The optimization problem can be written as

$$(4.3.21)\quad |\ \mathrm{diag}(\ (Y_c - XAA'X'Y_c)'(Y_c - XAA'X'Y_c)\ )\ | \rightarrow \min, \quad \text{with the constraint } (XA)'(XA) = I_r .$$

Using Lagrange multipliers, a general iterative optimization procedure could then be used to solve this maximization problem. A very natural alternative method, independently proposed by van der Leeden (1990) and Schmidli (1992) is motivated by the following reasoning : For given parameter A, the reduced rank regression model (4.3.7) is equivalent to a multivariate regression model with predictor matrix Z=XA. Hence for given A, the parameters B, Σ, μ can be easily estimated. On the other hand, if Σ is given, one has the case of a reduced rank regression model with known error covariance matrix and can use the analytical method derived in section 4.3.9.1 to estimate the parameter A. This leads to an iterative numerical approach, the Gauss–Seidel algorithm (see, for example, Thisted, 1988, pp.187), known under various other names (cyclic ascent, backfitting, alternating maximum likelihood, etc.), recently extended to ECM (see Meng and Rubin, 1993) and used in many statistical methods like GIFI (Gifi, 1990) and ACE (Breiman and Friedman, 1985). The Gauss–Seidel method can be summarized as follows. Suppose that a function $f(\theta)$ has to be maximized, and the parameter vector $\theta = (\theta_1 \ \theta_2)$ can be split in two parts such that for fixed θ_1 the optimal solution $\theta_2(\theta_1)$ of $f(\theta_1 \ \theta_2)$ is "easily" calculated and vice–verse. Then the Gauss–Seidel algorithm works as follows :

Let θ_1^0 be an initial estimate of the optimal value of θ_1
Then, until convergence, do for i = 1,2, ...
– $\theta_2^i := \theta_2(\theta_1^{i-1})$
– $\theta_1^i := \theta_1(\theta_2^i)$.

If $f(\theta)$ is a bounded function, then the Gauss–Seidel algorithm converges, although not necessarily to the globally optimal value.

This algorithm can be very naturally applied to the reduced rank regression models with diagonal error covariance matrix and is described in the appendix.

4.3.9. Asymptotic Distribution of the Predictions

To assess the predictive power of a prediction method, overall prediction criteria such as the mean squared error of prediction (MSEP) have been presented in section 2.3. One possibility to estimate these criteria is to use an asymptotic normal approximation to describe the variability of the predictor at a given point in prediction space (see section 2.3.5.1), which can be easily derived once the asymptotic distribution of the estimated coefficient matrix M is known.

For the case of a stochastic predictor matrix X, Izenman (1975) and later Velu et al. (1986) have derived the asymptotic distribution of the estimated coefficient matrix M and of the parameters A,B for the reduced rank regression model (4.3.7) with unstructured error covariance matrix (section 4.3.8.3), using perturbation methods. For the case of a fixed predictor matrix X, the asymptotic distribution of the maximum likelihood estimate of the coefficient matrix $M := AB'$ of the reduced rank regression model (4.3.7) can be calculated by first deriving the information matrix of the parameters A, B. From this information matrix, the asymptotic covariance of the estimates of the parameters A, B can be calculated and then, using the delta–method, the covariance matrix of their product. For the case of a coefficient matrix M of rank one, the asymptotic distribution of the estimated M has been given without proof by Jöreskog and Goldberger (1975). For the general case of fixed arbitrary rank, a sketch of the proof has been given by Ryan et al. (1992), however with an incorrect asymptotic covariance matrix for the estimated parameters A, B. A more detailed derivation of the asymptotic distribution with the correct asymptotic covariance matrix is now given. The results will later be used to derive the asymptotic MSEP for reduced rank regression models (see section 4.5.3).

Suppose that the sufficiently regular likelihood is given by $L(Y;\theta) \equiv p(Y|\theta)$ with parameter vector θ, and that the data Y consists of n observations which are generated from $p(Y|\theta_0)$, for some unknown θ_0. The maximum likelihood estimate $\hat{\theta}_n \equiv \hat{\theta}(Y)$ is the parameter value maximizing the likelihood. Let $l(Y;\theta) = \log L(Y;\theta)$ be the log–likelihood and

$$H(Y;\theta) = \partial^2 l(Y;\theta)/\partial\theta\partial\theta'$$

be the Hessian matrix of the log–likelihood. Then the Fisher–information matrix is defined as

$$\Omega_n(\theta_0) = -E_Y H(Y;\theta_0)$$

and the asymptotic Fisher–information matrix as

$$\Omega(\theta_0) = \lim 1/n \; \Omega_n(\theta_0) .$$

Then, the asymptotic covariance matrix of the maximum likelihood estimate is the inverse of the asymptotic Fisher–information matrix. More precisely,

$$\sqrt{n}\,(\hat{\theta}_n - \theta_0) \xrightarrow{D} N(0, \Omega(\theta_0)^{-1}) .$$

An estimate of the asymptotic covariance matrix of the maximum likelihood estimate is then, for example, given by

$$\Omega(\hat{\theta}_n)^{-1}.$$

For reduced rank regression, the asymptotic covariance matrix is now derived. Apart from an irrelevant constant, the log likelihood (4.3.8) $l \equiv l(Y; \Sigma, \mu, A, B)$ of the observed data given the parameters, can be written as

$$(4.3.22)\quad -2/n \cdot l = -\log|\Sigma^{-1}| + \mathrm{tr}[\Sigma^{-1}\ Z'Z]\ , \text{where}$$
$$Z = (Y - 1_n\mu' - XAB')/\sqrt{n}\,.$$

Using the notation of Magnus and Neudecker (1988), the first differential of l is given by

$$(4.3.23)\quad -2/n \cdot d\, l = \mathrm{tr}[(Z'Z - \Sigma)d\Sigma^{-1}] + \mathrm{tr}[\Sigma^{-1}\ dZ'Z]\ ,$$
since
$$d\log|\Sigma^{-1}| = \mathrm{tr}[\Sigma d\Sigma^{-1}] \quad \text{and}$$
$$d\ \mathrm{tr}[\Sigma^{-1}\ Z'Z] = \mathrm{tr}[d\Sigma^{-1}\ Z'Z] + \mathrm{tr}[\Sigma^{-1}\ dZ'Z]\,.$$

The second differential of l is given by

$$(4.3.24)\quad -2/n \cdot d^2\, l$$
$$= \mathrm{tr}[(dZ'Z - d(\Sigma^{-1})^{-1})d\Sigma^{-1}] + \mathrm{tr}[(Z'Z - \Sigma)d^2\Sigma^{-1}]$$
$$+ \mathrm{tr}[d\Sigma^{-1}\ dZ'Z] + \mathrm{tr}[\Sigma^{-1}\ d^2Z'Z]$$
$$= \mathrm{tr}[(2Z'dZ + \Sigma d(\Sigma^{-1})\Sigma)d\Sigma^{-1}]$$
$$+ \mathrm{tr}[(Z'Z - \Sigma)d^2\Sigma^{-1}] + \mathrm{tr}[d\Sigma^{-1}\ 2Z'dZ]$$
$$+ \mathrm{tr}[\Sigma^{-1}\ 2(Z'd^2Z + (dZ)(dZ)')]\,,$$
since $d((\Sigma^{-1})^{-1}) = \Sigma\ d(\Sigma^{-1})\ \Sigma$.

Taking expectations, one obtains

$$(4.3.25)\quad -2/n \cdot E\ d^2\, l$$
$$= \mathrm{tr}[\Sigma\ d\Sigma^{-1}\ \Sigma\ d\Sigma^{-1}] + E\ 2\ \mathrm{tr}[(dZ)'\Sigma^{-1}(dZ)]\ ,$$
since $E\ Z = 0$ and $E\ Z'Z = \Sigma$.

Using vector notation, one can write the first part of (4.3.25) as

$$(4.3.26)\quad \mathrm{tr}[\Sigma\ d\Sigma^{-1}\ \Sigma\ d\Sigma^{-1}]$$
$$= \mathrm{vec}(d\Sigma^{-1})'\ (\Sigma \otimes \Sigma)\ \mathrm{vec}(d\Sigma^{-1})$$
$$= 2\ v(d\Sigma^{-1})'\ \Omega_{00}\ v(d\Sigma^{-1})\ , \text{where}$$
$$\Omega_{00} = 1/2\ D'(\Sigma \otimes \Sigma)D\ , \text{with}$$
D the duplication matrix and v(.) the vectorization of the unique elements of a symmetric matrix (see Magnus and Neudecker, 1988, pp.48).

Note that if it is known that Σ is diagonal, then v(.) is the vectorization of the diagonal elements of a matrix; if it is known that Σ is proportional to the identity matrix, then v(.) is equal to the proportionality factor. In both cases, the matrix D has to be adapted.

The second part of (4.3.25) can be written as

(4.3.27) $$2 \ E \ tr[(dZ)'\Sigma^{-1}(dZ)]$$
$$= 2/n \ tr[(d(1_n\mu' + XAB'))'\Sigma^{-1}(d(1_n\mu' + XAB'))]$$
$$= 2/n \ vec(d(1_n\mu' + XAB'))' (\Sigma^{-1} \otimes I_n) \ vec(d(1_n\mu' + XAB'))$$
$$= 2 \ (vec(d\mu')' \ vec(A)' \ vec(B')') \ \Omega \ (vec(d\mu')' \ vec(A)' \ vec(B')')',$$

with

$$\Omega = \begin{bmatrix} \Omega_{11} & 0 & 0 \\ 0 & \Omega_{22} & \Omega_{23} \\ 0 & \Omega_{32} & \Omega_{33} \end{bmatrix}, \text{ where}$$

$$\Omega_{11} = 1/n \ \Sigma^{-1} \otimes 1_n'1_n = \Sigma^{-1},$$
$$\Omega_{22} = 1/n \ (B \otimes X)'(\Sigma^{-1} \otimes I_n)(B \otimes X) = B'\Sigma^{-1}B \otimes (X'X/n)$$
$$\Omega_{33} = 1/n \ (I_q \otimes XA)'(\Sigma^{-1} \otimes I_n)(I_q \otimes XA) = \Sigma^{-1} \otimes A'(X'X/n)A$$
$$\Omega_{23} = \Omega_{32}' = 1/n \ (B \otimes X)'(\Sigma^{-1} \otimes I_n)(I_q \otimes XA) = B'\Sigma^{-1} \otimes (X'X/n)A, \text{ since}$$

$vec(dXAB') = (B \otimes X)vec(dA) + (I_q \otimes XA)vec(dB')$
and $1_n'X = 0$.

Hence, the information matrix with respect to the parameters $(\Sigma^{-1}, \mu, A, B')$ is given by

(4.3.28*) $$\Omega_n = n \begin{bmatrix} \Omega_{00} & 0 & 0 & 0 \\ 0 & \Omega_{11} & 0 & 0 \\ 0 & 0 & \Omega_{22} & \Omega_{23} \\ 0 & 0 & \Omega_{32} & \Omega_{33} \end{bmatrix}.$$

Note, that this information matrix is singular.

For the moment suppose that a specific generalized inverse of the asymptotic information matrix $\Omega := \lim 1/n \ \Omega_n$ gives the asymptotic covariance matrix of the maximum–likelihood estimates of $(\Sigma^{-1}, \mu, A, B')$, as claimed by Ryan et al. (1992), a point that will be discussed below. The used generalized inverse is given by

$$(4.3.29^*) \quad \Omega^- = \begin{bmatrix} \Omega_{00}^{-1} & 0 & 0 & 0 \\ 0 & \Omega_{11}^{-1} & 0 & 0 \\ 0 & 0 & \tilde{\Omega}_{22} & 0 \\ 0 & 0 & 0 & \tilde{\Omega}_{33} \end{bmatrix} \quad \text{, where}$$

$$\begin{aligned} \Omega_{00}^{-1} &= 2\, D^{+\prime}(\Sigma^{-1} \otimes \Sigma^{-1})D^{+} \ , \\ \Omega_{11}^{-1} &= \Sigma \ , \\ \tilde{\Omega}_{22} &= (B'\Sigma^{-1}B)^{-1} \otimes (Q^{-1} - A(A'QA)^{-1}A') \ , \\ \tilde{\Omega}_{33} &= \Sigma \otimes (A'QA)^{-1} \quad \text{and} \\ Q &= \lim\, (X'X/n) \text{ nonsingular} \ . \end{aligned}$$

Still supposing that Ω^- is in fact the asymptotic covariance matrix of the maximum–likelihood estimates, one can derive the asymptotic covariance matrix of the coefficient matrix $\hat{A}\hat{B}'$ as

$$(4.3.30^*) \quad n\ \mathrm{cov}(\hat{A}\hat{B}') \rightarrow G \begin{pmatrix} \tilde{\Omega}_{22} & 0 \\ 0 & \tilde{\Omega}_{33} \end{pmatrix} G' \quad \text{, where}$$

$$G := (\ (B \otimes I_p) \ \ (I_q \otimes A)\ ).$$

From this, one obtains

$$\begin{aligned} &(4.3.31^*) \quad \mathrm{cov}(\hat{A}\hat{B}') \\ &\approx B(B'\Sigma^{-1}B)^{-1}B' \otimes ((X'X)^{-1} - A(A'(X'X)A)^{-1}A') \\ &\qquad\qquad + \Sigma \otimes A(A'X'XA)^{-1}A' \\ &= \Sigma \otimes (X'X)^{-1} \\ &\quad - (\Sigma - B(B'\Sigma^{-1}B)^{-1}B') \otimes ((X'X)^{-1} - A(A'(X'X)A)^{-1}A'). \end{aligned}$$

First note that the asymptotic covariance matrix of the estimated coefficient matrix is invariant under transformations the parameters $(A,B) \rightarrow (AF, BF^{-1})$. Further note that from the second way of writing the covariance matrix in (4.3.31*), it is obvious, that reduced rank regression is asymptotically more efficient than multivariate regression, since the covariance matrix of the estimated coefficient matrix obtained by multivariate regression is given by

$$\Sigma \otimes (X'X)^{-1} .$$

A proper solution to the problem of the singular information matrix (4.3.28*) is now given, deriving the correct asymptotic covariance of the estimated parameters A, B. The reason for the singularity of the information matrix is the non–uniqueness of the parameters A, B. Hence a different parametrization of A is necessary to be able to properly derive the asymptotic covariance matrix of the new estimated parameters. For the present purpose it is more convenient not to use the standard parametrization III (4.3.6) of A. Instead, without loss of generality (perhaps after a re–ordering of the columns of the predictor matrix X), it is assumed that the parameter A is of the form corresponding to parametrization I (4.3.4), namely

(4.3.28) $A = \begin{pmatrix} I_r \\ \theta' \end{pmatrix}$, with arbitrary $(p-r) \times r$ matrix θ'.

Then one has

(4.3.29) $\mathrm{vec}(dA) = K_{p,r} \begin{pmatrix} 0_{r^2 \times r(p-r)} \\ I_{r(p-r)} \end{pmatrix} \mathrm{vec}(d\theta) \quad =: H\ \mathrm{vec}(d\theta)$,

since
$\mathrm{vec}(A) = K_{p,r}\ (\mathrm{vec}(I_r)'\ \ \mathrm{vec}(\theta)')'$, where
$K_{p,r}$ is the commutation matrix (see Magnus and Neudecker, 1988, p.46).

Hence, the asymptotic information matrix with respect to the new parameters $(\Sigma^{-1}, \mu, \theta, B')$ is given by

(4.3.30) $\Omega = \begin{bmatrix} \Omega_{00} & 0 & 0 & 0 \\ 0 & \Omega_{11} & 0 & 0 \\ 0 & 0 & H'\Omega_{22}H & H'\Omega_{23} \\ 0 & 0 & \Omega_{32}H & \Omega_{33} \end{bmatrix}$.

Since Ω is not singular, its inverse Ω^{-1} is the asymptotic covariance matrix of the maximum–likelihood estimates with respect to the parameters $(\Sigma^{-1}, \mu, \theta, B')$. Note that the maximum likelihood estimates of θ and B are asymptotically correlated (compare with (4.3.29*)). The asymptotic covariance matrix of the coefficient matrix $\hat{A}\hat{B}'$ is now given by

(4.3.31) $n\ \mathrm{cov}(\hat{A}\hat{B}') \rightarrow G\tilde{H}(\tilde{H}'\tilde{\Omega}\tilde{H})^{-1}\tilde{H}'G'$, where

$\tilde{\Omega} := \begin{pmatrix} \Omega_{22} & \Omega_{23} \\ \Omega_{32} & \Omega_{33} \end{pmatrix}$,

$G := (\ (B \otimes I_p)\ \ (I_q \otimes A)\ )$ and

$\tilde{H} := (H'\ \ I_{qr})'$.

This is identical to (4.3.31*), which can be seen as follows. First,

$\tilde{H}(\tilde{H}'\tilde{\Omega}\tilde{H})^{-1}\tilde{H}'$

is a generalized inverse of $\tilde{\Omega}$. If one can show that $G\tilde{\Omega}^{-}G'$ is independent of the choice of the generalized inverse, then the generalized inverse defined in (4.3.29*) can be used. Note that one can write

$$(4.3.32)\quad \tilde{\Omega} = V'V \text{ , where}$$
$$V' := G' \quad (\Sigma^{-1/2} \otimes (X'X)^{1/2}) \ .$$

Then, for any generalized inverse $\tilde{\Omega}^{-}$ of $\tilde{\Omega}$, $V\tilde{\Omega}^{-}V'$ is independent of the choice of the generalized inverse. But

$$V\tilde{\Omega}^{-}V' = (\Sigma^{-1/2} \otimes (X'X)^{1/2})\, G\tilde{\Omega}^{-}G'(\Sigma^{-1/2} \otimes (X'X)^{1/2}),$$

and one has the required result.

An estimate of the asymptotic covariance matrix of the maximum likelihood estimates of $(\Sigma^{-1}, \mu, \theta, B')$ is obtained from the inverse of (4.3.30) by replacing the true values of the parameters by their maximum likelihood estimates. An estimate of the asymptotic covariance matrix of the estimated M is obtained from either (4.3.31) or (4.3.31*), by replacing the true values of the parameters A and B by their estimates.

If the variability of the predictor at a given point x in the predictor space is of interest, the asymptotic covariance of the estimated coefficient matrix M can be used as follows. First note that the predicted value is given by

$$\hat{y} = \hat{\mu} + (x'\hat{M})' = \hat{\mu} + (I_q \otimes x')\text{vec}(\hat{M}).$$

Hence,

$$\text{cov}(\hat{y}) = \Sigma/n + (I_q \otimes x')\text{cov}(\hat{M})(I_q \otimes x')'.$$

Using (4.3.31*) one obtains

$$(4.3.33)\quad \text{cov}(\hat{y}) \approx \Sigma/n + \Sigma \cdot x'(X'X)^{-1}x$$
$$- (\Sigma - B(B'\Sigma^{-1}B)^{-1}B') \cdot$$
$$(x'(X'X)^{-1}x - x'A(A'X'XA)^{-1}A'x).$$

To obtain some insight about how much in precision is actually gained by using reduced rank regression instead of multivariate regression, it is instructive to consider the case of a reduced rank regression model of rank r whose error covariance matrix is proportional to the identity matrix.

Measuring the size of a covariance by its trace, the ratio of the covariance sizes for reduced rank regression and multivariate regression is given for large sample sizes by

$$R := \mathrm{tr}\{ \mathrm{cov}(\hat{y}) \} / \mathrm{tr}\{ \Omega \}$$
$$= r/q + (1 - r/q)\ x'AA'x/\{ x'(X'X)^{-1}x\} ,$$

recalling that $A'X'XA = I_r$. Hence, if the point x in predictor space lies in the subspace spanned by the latent variables, the ratio R is equal to 1 and nothing in precision is gained. However if x is orthogonal to the latent factor subspace, the ratio R reduces to r/q, so that especially for small ranks r and for a large number of responses q, the gains in precision can be considerable.

The situation is different, when tolerance regions for the "future" response y are considered. Then formally a term Σ has to be added to the covariance of the predictor and drastic differences between multivariate regression and reduced rank regression can only be expected for points x in predictor space with

$$1 \leq x'(X'X)^{-1}x ,$$

that is for extrapolation.

An approximate tolerance region for a future observation can be based on a normal approximation with covariance matrix (4.3.33). There the unknown parameters have to be replaced by their maximum likelihood estimates and the measurement error has to be considered as well (addition of an estimate of Σ). An approximate p%–tolerance region is then defined by

$$\text{(4.3.34)}\quad (y - \hat{y})'\{\hat{\Sigma} + \hat{\mathrm{cov}}(\hat{y})\}^{-1}(y - \hat{y}) \leq \chi_n^2(p) ,$$

where $\chi_n^2(p)$ is the quantile of the chi–square distribution with n degrees of freedom (see, for example, Mardia et al., 1979).

Note that the tolerance region is based on asymptotic theory and its goodness for small samples is unknown. In contrast to this ad hoc procedure, a Bayesian analysis of reduced rank regression is presented in chapter 5, using the Gibbs sampling procedure (see, for example, Gelfand and Smith, 1990). This allows to calculate the predictive density almost exactly, even for small sample sizes.

4.3.10. Example

For the illustrative QSAR example presented in section 1.6, a reduced rank regression model (4.3.7) of rank 1 with arbitrary error covariance matrix seems to be adequate (CCA, canonical correlation analysis, see section 4.3.8.3). The maximum likelihood estimates are given in Table 4.3.1.

Table 4.3.1 Maximum likelihood estimates and asymptotic standard deviations

μ	ADULTS	LARVAE
	-1.68	-1.75

Σ	ADULTS	LARVAE
ADULTS	0.0589	0.0277
LARVAE	0.0277	0.0787

A	Coeff.	Std.Dev
I1	1.000	0.000
I2	-1.199	0.230
IP23	-0.049	0.248
Pi	-0.675	0.575
F	0.201	0.263
R	0.561	0.386
MR	-0.598	0.510
L	-0.118	0.651
B1	0.911	0.759

B	ADULTS	LARVAE
Coeff.	-0.2677	-0.2560
Std.Dev.	0.0495	0.0510

Comparison of reduced rank regression (CCA) and multivariate regression (MR).

CCA	coeff.		Std.Dev.		MR coeff.		Std.Dev.	
	ADULTS	LARVAE	ADULTS	LARVAE	ADULTS	LARVAE	ADULTS	LARVAE
I1	-0.267	-0.256	0.049	0.051	-0.283	-0.215	0.051	0.054
I2	0.321	0.307	0.052	0.055	0.308	0.338	0.054	0.057
IP23	0.013	0.012	0.068	0.065	-0.041	0.150	0.073	0.077
Pi	0.180	0.173	0.155	0.149	0.236	0.031	0.166	0.175
F	-0.054	-0.051	0.072	0.069	-0.035	-0.098	0.077	0.081
R	-0.150	-0.143	0.098	0.094	-0.109	-0.246	0.106	0.111
MR	0.160	0.153	0.135	0.129	0.064	0.396	0.145	0.152
L	0.031	0.030	0.177	0.169	0.117	-0.186	0.191	0.200
B1	-0.244	-0.233	0.206	0.198	-0.262	-0.187	0.222	0.233

For the estimates of the parameters A, B and M, their asymptotic standard deviations are calculated according to section 4.3.9. They are shown as well as the standard deviations for the estimated M which are obtained from an ordinary multivariate regression (MR, see section 4.3.1). Note that since for the asymptotic estimates the non–standard parametrization I (4.3.4) of the matrix A is needed (as explained in section 4.3.9), the coefficient of the variable I1 in the coefficient matrix A is set to 1.

From the estimates of μ and B, one sees that there are almost no shifting and scaling differences between the two response variables. The estimate of the error covariance matrix Σ suggests that the errors are correlated, which is expected since the two responses are measured in the same experiment. Comparing the estimate of A and its asymptotic standard deviation, it seems that besides I1, the variables I2, F, R and perhaps MR are the most important variables to explain the variation of the responses. However, since the predictors are correlated, this could be misleading. A comparison of the estimated standard deviations of the coefficient matrix M obtained by reduced rank regression and by ordinary multivariate regression shows that the rank restriction results at least asymptotically in more precise parameter estimates.

Table 4.3.2 Predictions and approximate 90% tolerance intervals for the test data using reduced rank regression with arbitrary error covariance matrix (Canonical Correlation Analysis, CCA, see section 4.3.8.3)). The two outlier substances are shown in bold face (compare with section 3.5)

	ADULTS				LARVAE			
			tol.interv.				tol.interv.	
	obs.	pred.	upper	lower	obs.	pred.	upper	lower
IIca	-2.0	-1.80	-2.43	-1.18	>-1.7	-1.86	-2.54	-1.19
IIcb	>-1.7	-2.35	-3.01	-1.68	-2.0	-2.38	-3.10	-1.67
IIia	**-2.6**	**-0.93**	**-1.99**	**0.12**	**-2.6**	**-1.03**	**-2.10**	**0.02**
IIIia	**>-1.7**	**-0.30**	**-1.33**	**0.71**	**-2.0**	**-0.43**	**-1.48**	**0.61**
IIIka	>-1.7	-0.93	-1.56	-0.31	-1.7	-1.03	-1.72	-0.35
IIIla	>-1.7	-1.04	-1.59	-0.48	>-1.7	-1.13	-1.75	-0.51
IIIma	>-1.7	-1.06	-1.74	-0.39	>-1.7	-1.16	-1.88	-0.44

A comparison with the coefficient matrices obtained by principal component regression (PCR) and partial least squares (PLS) in Table 3.5.3 shows quite large differences for some coefficients.

In Table 4.3.2 the predicted values for the test dataset are shown as well as the approximate 90% tolerance intervals based on asymptotic normal approximations presented in section 4.3.9. One sees again, that substances IIia and IIIia are clear outliers. As noted in section 3.5, these two substances are unstable under UV–light (Pascual et al., 1994). The predictions are relatively similar to those obtained by principal component regression PCR and partial least squares PLS (compare again with Table 3.5.3), although, as noted above, this is not the case for the coefficient matrices. This shows again a well known problem of regression methods : if the predictor variables are correlated, great care is necessary when the coefficients are interpreted.

4.4. Extensions of the Standard Reduced Rank Regression Model

The reduced rank regression model (4.3.7) considered in the preceding section 4.3 can be extended in various ways, some of which will be shortly discussed; their application is beyond the scope of this treatise and these methods will not be used later on.

4.4.1. Structured Error Covariance Matrix

If the measurement procedure is known in very great detail, it is possible to incorporate this knowledge by assuming that the error covariance matrix Σ in (4.3.7) can be written as $\Sigma = \Sigma(\theta)$ for some parameter vector θ, which can be independently varied from the other parameters in model (4.3.7). This setting has been broadly discussed in vanderLeeden (1990). The four different structures of the error covariance matrix presented in the section 4.3.6 are special cases of this more general setting. However while for these the maximum likelihood estimates can be obtained analytically, with the exception of the case of a diagonal error covariance matrix, only numerical solutions are available for this general setting. A Gauss–Seidel algorithm is proposed for optimization (called alternating maximum likelihood), already presented in section 4.3.9.4. However, to find the maximum likelihood estimate of the parameter θ for given parameter A, a numerical optimization is usually necessary, for example the Fletcher–Powel algorithm.

In QSAR studies it is usually the case that the most adequate structure among the four different error covariance structures presented in section 4.3.9 can be identified, and a more detailed error structure is usually not available.

4.4.2. Latent Variable Models

Reduced rank regression models (4.3.7) can be motivated by the latent variable concept: the expected value of the response matrix Y equals a linear transformation of the latent variable matrix Z, which in turn is a linear transformation of the predictor matrix X (see section 4.2). Jöreskog and Goldberger (1975) extended this concept by assuming that Z is a random variable, whose expected value is a linear transformation of the predictor matrix X. The resulting model is called a MIMIC model (Multiple Indicators and Multiple Causes) with one–dimensional latent variable Z. A slight generalization of the MIMIC model, using the same notation as in the preceding section can be written as

(4.4.1) $Y \mid Z \sim N(1_n\mu' + ZB', \Sigma \otimes I_n)$,
$Z \sim N(XA, \Omega \otimes I_n)$, where
Ω is a positive definite $r \times r$ matrix of unknown parameters.

Note that by substitution the model can be rewritten as

(4.4.2) $Y \sim N(1_n\mu' + XAB', (B\Omega B' + \Sigma) \otimes I_n)$.

The maximum likelihood estimates of the parameters have to be obtained numerically (Jöreskog and Goldberger, 1975).

In the area of QSAR, it is not obvious how a MIMIC model could be interpreted. One possible interpretation is that there are random environmental variables which affect the latent variable. For example, if herbicides are tested in the glasshouse, the weather may influence the biological effect of a substance, i.e. the latent variable.

4.4.3. Non–normal Errors, Outliers and Robustification

Up to know, it has been assumed that the errors in the reduced rank regression model (4.3.7) are normally distributed. However, as seen in the derivation of the maximum likelihood estimates, these are equivalent to weighted least squares procedures with the exception of the case of a diagonal error covariance matrix. Therefore, the same estimates can be used for non–normal errors, if the least squares criterion seems adequate. If it is suspected that outliers may occur, then robust versions of reduced rank regression should be considered. One possibility is to use the L1–norm instead of the euclidean norm in the weighted least squares estimation procedures.

4.4.4. Nonlinearities

Reduced rank regression models have been motivated by the application area of QSAR in section 4.2. It was already noted, that the linear relationship between the biological response variables and the latent variables as well as the linear relationship between the latent variables and the physico–chemical predictor variables is an approximation. A general nonlinear reduced rank regression model can be written as

(4.4.3) $y = f(g(x)) + e$, where
e is a q–dimensional random variable of errors and
$g : R^p \rightarrow R^r$ and $f : R^r \rightarrow R^q$ are functions.

For the univariate case, that is for the case of just one response variable, many nonparametric methods are available to estimate the relationship between response variable and predictor variables, for example ACE (Breiman and Friedman, 1985). Nonparametric methods to model multivariate responses are rarely considered. One main exception seems to be the GIFI methodology (Gifi, 1990), which may be adapted to produce models similar to (4.4.3). Another exception is multivariate projection pursuit regression (Friedman, 1985), which can be regarded as a nonlinear generalization of redundancy analysis (section 4.3.8.2), since the following model is implicitly estimated

(4.4.4) $y = f(x'A)B' + e$, where
e is a q–dimensional random variable of errors with covariance matrix proportional to the identity matrix,
$f : R^r \rightarrow R^r$ which transform each latent variable $z' = x'A$ separately, and
A and B are $p \times r$ and $q \times r$ matrices of parameters.

The main problem with the application of nonparametric methods is their instability. If the number of observations is relatively high and the errors are small compared with the deviations from linearity, then these methods are certainly useful. However, if their variability is much larger than the bias introduced by using linear methods, then linear methods should clearly be preferred. In QSAR studies, only few observations are usually available compared with the number of physico–chemical predictor variables. Additionally, the errors are relatively large compared with the variation of the response variables. Hence, rather than use a nonparametric approach to estimate nonlinearities, occasionally transformations of some of the predictor variables like squares or interactions will be used to augment the set of predictor variables. Obviously, the linear reduced rank regression model can still be used in these cases.

4.4.5. Econometric Models

In econometrics, it is typically the case that variables are observed over time. When one set of such variables is to be predicted by another set of variables, a multivariate regression model is often adequate. However, in contrast to the models (4.3.1) or (4.3.7), the errors are usually correlated for different observations (time points). Then the covariance of the error matrix E in (4.3.7) cannot be written as $\Sigma \otimes I_n$ but rather as $\Sigma \otimes \Omega(\theta)$, where $\Omega(\theta)$ is a structured error covariance matrix catching the correlation structure over time. In general, the maximum likelihood estimates cannot be obtained analytically.

Another use of reduced rank regression models is the analysis of multivariate time–series. Suppose that the q–dimensional vector y_t is observed at time t = 1, ... , n. Then an autoregressive model of order k can be written as

$$y_t = \sum_{i=1}^{k} M_i y_{t-i} + \epsilon_t ,$$

or in matrix form as $Y = X M + E$, where X contains the lagged variables $y_{t-1}, ..., y_{t-k}$ and E is the error matrix. Assuming that M is not of full rank, a reduced rank regression model (4.3.7) is obtained (see Velu et al., 1986). This model can be refined by considering additional structure in the coefficient matrix M, for example corresponding to a nested autoregressive model (see Ahn and Reinsel, 1988). Another application of reduced rank regression models for multivariate time–series is in the context of cointegration (see, for example, Geweke, 1993).

Economic systems can sometimes be modeled by simultaneous equation systems, linking q endogenous variables Y and p exogenous variables X (see, for example, Mardia et al., 1979, chapter 7, or Toutenburg, 1992, pp.3). A simultaneous equation system can be written as

$YB + XA = U,$

where U is an error matrix, B is a nonsingular $q \times q$ coefficient matrix and A is a $p \times q$ coefficient matrix. Multiplying the simultaneous equation by B^{-1}, the so called reduced form is obtained as

$Y = X M + E,$

where $M = AB^{-1}$ and $E = UB^{-1}$, that is a multivariate regression model (4.3.1).

As shown by Geweke (1993), a reduced rank regression model is usually obtained, when concentrating on only one equation in the original simultaneous equation system.

4.5. Prediction Criteria for the Rank Selection of Reduced Rank Regression Models

In reduced rank regression, it is assumed that the coefficient matrix M of a multivariate regression model is not of full rank. If the rank is not exactly known a–priori, criteria for the selection of the rank have to be used. In a setting where the interest centers on prediction, the selection criterion should reflect the predictive power of the model. Two possible prediction criteria introduced in section 2.3, the information criterion IC and the mean squared error of prediction MSEP are discussed in section 4.5.2 and 4.5.3, respectively, where their ability to select the correct rank is investigated by a simulation study. Finally an illustrative QSAR example is presented in section 4.5.4.

4.5.1. Likelihood

Suppose that an $n \times q$ response matrix Y is the realization of a random variable following the reduced rank regression model (4.3.7), with the average log likelihood given by (4.3.8). The maximum likelihood estimates of the reduced rank regression model have been given in section 4.3 for different structures of the error covariance matrix. The three most relevant cases will be discussed in the following, namely the case of an unstructured error covariance matrix (CCA), of an error covariance matrix proportional to the identity matrix (RDA) and of a diagonal error covariance matrix (RRR); see section 4.3.6. For fixed A, B, μ, the maximum likelihood estimate of Σ is a simple function of $S(Y; A, B, \mu)$ defined by (4.3.8). Its estimates and the logarithm of its determinant will be used in following section 4.5.2 to calculate the information criterion IC. They are given by

(4.5.1a) Σ arbitrary (canonical correlation analysis, CCA, see section 4.3.8.3) :

$$\hat{\Sigma}(Y; A, B, \mu) = S(Y; A, B, \mu) \text{ and}$$

$$\log |\hat{\Sigma}(Y; A, B, \mu)| = \sum_{i=1}^{q} \log(\lambda_i^2) \text{ , where}$$

$S(Y; A, B, \mu) = V \Lambda V'$ is the singular value decomposition with $\Lambda = \mathrm{Diag}(\lambda_1^2, ..., \lambda_q^2)$ diagonal;

(4.5.1b) $\Sigma = \mathrm{Diag}(\sigma_1^2, ..., \sigma_q^2)$ (reduced rank regression with diagonal error covariance matrix, RRR, see section 4.3.8.4) :

$$\hat{\Sigma}(Y; A, B, \mu) = \mathrm{Diag}(s_1^2, ..., s_q^2) \text{ , where}$$

$$s_i^2 := (S(Y; A, B, \mu))_{ii} \text{ , and}$$

$$\log |\hat{\Sigma}(Y; A, B, \mu)| = \sum_{i=1}^{q} \log(s_i^2) \text{ ;}$$

(4.5.1c) $\Sigma = \sigma^2 I$ (redundancy analysis, RDA, see section 4.3.8.2) :

$$\hat{\Sigma}(Y; A, B, \mu) = s^2 I_q \text{ , where}$$

$$s^2 := 1/q \sum_{i=1}^{q} (S(Y; A, B, \mu))_{ii} \text{ , and}$$

$$\log |\hat{\Sigma}(Y; A, B, \mu)| = q \; \log(s^2) \text{ .}$$

If these estimates are put into the average log likelihood (4.3.8), one obtains the concentrated likelihood

(4.5.2) $l(Y; A, B, \mu)$

$$= \log |\hat{\Sigma}(Y; A, B, \mu)| + \mathrm{tr}[S(Y; A, B, \mu) \hat{\Sigma}^{-1}(Y; A, B, \mu)]$$

$$= \log |\hat{\Sigma}(Y; A, B, \mu)| + q.$$

Note that the second term is identical for all three different error covariance structures. Replacing the parameters A, B, μ by their maximum likelihood estimates $\hat{A}(Y), \hat{B}(Y), \hat{\mu}(Y)$, the estimated error covariance matrix is given by

$$\hat{\Sigma}(Y) := \hat{\Sigma}(Y; \hat{A}(Y), \hat{B}(Y), \hat{\mu}(Y)).$$

4.5.2. Information Criterion for Rank Selection

4.5.2.1. Estimation of the Information Criterion

The information criterion IC has been introduced in section 2.3 as a measure of the predictive power of a density predictor. For reduced rank regression models (4.3.7), the information criterion (2.3.8) is defined as

(4.5.3) $IC := -E_{Y|X} E_{\tilde{Y}|X} l(\tilde{Y}; \hat{A}(Y), \hat{B}(Y), \hat{\Sigma}(Y), \hat{\mu}(Y))$, where $\tilde{Y}|X$ is independent and identically distributed as $Y|X$.

The fit, asymptotic and crossvalidation estimates introduced in section 2.3.5.2 are obtained as follows.

(4.5.4) $IC_{FIT} := -q - \log|\hat{\Sigma}(Y; \hat{A}(Y), \hat{B}(Y), \hat{\Sigma}(Y), \hat{\mu}(Y)|$.

(4.5.5) IC_{ASY}
$:= -q - \log|\hat{\Sigma}(Y; \hat{A}(Y), \hat{B}(Y), \hat{\Sigma}(Y), \hat{\mu}(Y)| + m(r)/n$,
where $m(r)$ is the number of free parameters in model (4.3.7) with rank r.

For the three different structures of the error covariance matrix (see section 4.5.1), the number of free parameters $m(r)$ is given by

(4.5.6) a) Σ arbitrary :
$m(r) = q + p \cdot r + q \cdot r - r^2 + q(q+1)/2;$

b) $\Sigma = \mathrm{Diag}(\sigma_1^2, ..., \sigma_q^2)$:
$m(r) = q + p \cdot r + q \cdot r - r^2 + q;$

c) $\Sigma = \sigma^2 I$: $m(r) = q + p \cdot r + q \cdot r - r^2 + 1$.

To obtain the crossvalidation estimate of the information criterion, define the initial crossvalidation estimate of the error covariance matrix as follows.

(4.5.7) $\hat{\Sigma}^*_{CV} := 1/n \sum_{i=1}^{n} (Y_{i.} - \hat{Y}_{(.)i.})'(Y_{i.} - \hat{Y}_{(.)i.})$.

Then the crossvalidation estimate of IC is given by

(4.5.8) $IC_{CV} := -q - \log|\hat{\Sigma}_{CV}|$, where

a) Σ arbitrary : $\hat{\Sigma}_{CV} := \hat{\Sigma}^*_{CV}$;

b) $\Sigma = \mathrm{Diag}(\sigma_1^2, ..., \sigma_q^2)$: $\hat{\Sigma}_{CV} := \mathrm{Diag}(s^2_{CV1}, ..., s^2_{CVq})$,

with $s^2_{CVi} := (\hat{\Sigma}^*_{CV})_{ii}$;

c) $\Sigma = \sigma^2 I$: $\hat{\Sigma}_{CV} = s^2_{CV} I_q$,

with $s^2_{CV} := 1/q \sum_{i=1}^{q} (\hat{\Sigma}^*_{CV})_{ii}$.

4.5.2.2. A Simulation Study

If the information criterion is used for rank selection, it is natural to select the rank with minimal estimated information criterion, that is with $r_0 := \arg\min IC(r)$, where $IC(r)$ is an estimate of the information criterion for fixed rank r. The fit estimate given in the preceding section 4.5.2.1 cannot be used for this rank selection procedure since otherwise always the model with maximal rank is selected.

To see how well the asymptotic estimate (4.5.5) and the crossvalidation estimate (4.5.8) of the information criterion IC are able to select the correct rank in practice, their behavior for small sample sizes should be known. Ideally their reaction to misspecifications like non–normal errors, wrong error structures, etc. should be investigated as well. However since it is not possible to provide analytical answers to these problems, a small simulation study is intended to give some insight into the goodness of the selection procedures. In the following, it will be assumed that the reduced rank regression model (4.3.7) is correct, i.e. that the only uncertainty about the model is the rank of the coefficient matrix. The information criterion IC depends on the predictor matrix X, the unknown parameters and the sample size n , so that for any setting of these parameters, one or the other rank selection method could be better. However for reduced rank regression models one can assume for the simulation, without loss of generality, that $\mu = 0$ and $\Sigma = I_q$, since the estimates of the information criterion are identical up to an additive constant under affine transformations of the responses.

For the simulation, the number of predictor variables is fixed to $p = 10$, and the number of response variables to $q = 5$. The true rank and the sample size are varied as follows.

- true rank $r = 1, 2, 3, 4, 5$;
- sample size $n = 25, 50, 75$.

For fixed true rank and sample size, the predictor matrix and the parameters are set as follows.

- $n \times p$ predictor matrix $X := (\tilde{X} - 1/n \; 1_n 1_n' \tilde{X})$, where $\tilde{X} \sim N(0, I_{np})$;
- $p \times r$ parameter matrix $A = (I_r \,:\, 0_{r \times (p-r)})'$;
- $q \times r$ parameter matrix : $B := f\tilde{B}$, where $\tilde{B} = (I_r \,:\, (1/r) \cdot 1_r 1_{q-r}')'$ and $f := \sqrt{1/((1 - R^2) \, \| XA\tilde{B}' \|^2 / nq)}$ with $R^2 := 0.8$.

The factor f is chosen to ensure a reasonable predictive power of the models characterized by the mean R^2 of 0.8.

For each fixed setting of the true rank and the sample size, the simulation proceeds as follows. Let N be a large integer, N=100, say, then for i=1, 2, ... , N,

sample $Y^{(i)} \sim N(XAB', I_{nq})$,
for assumed ranks j=1, 2, ... , q , calculate
$IC^{(i)}_{ASY}(j)$ and $IC^{(i)}_{CV}(j)$ as in (4.5.5), (4.5.8).

To judge the goodness of the two selection procedures, their ability to select the correct rank is investigated. In the part of Table 4.5.1 referring to the information criterion IC, the proportions of correct selections for given true rank are shown for the different settings of true rank and sample size. One sees that a too small rank is never selected, however higher than necessary ranks are obtained quite often, especially if the sample size is small. The criterion based on asymptotic theory is much better in selecting the correct rank than the crossvalidation criterion. The higher variability of the crossvalidation estimates in comparison with the asymptotic estimates has been noted in other simulation studies as well, mainly in connection with multiple regression (see, for example, Breiman and Spector, 1992).

Table 4.5.1 Simulation : Percentages of rank selection

For given method (RDA, CCA, RRR), sample size (n=25,50,75) and true rank (1,2,3,4,5), the dimension (rank) is selected by four different prediction criteria :

MSEPasy : asymptotic estimate of the mean squared error of prediction,
MSEPcv : crossvalidation estimate of the mean squared error of prediction,
ICasy : asymptotic estimate of the information criterion,
ICcv : crossvalidation estimate of the information criterion.

The results with respect to the criteria ICasy and ICcv are discussed in section 4.5.2.2, while the results with respect to the criteria MSEPasy and MSEPcv are discussed in section 4.5.3.2.

The percentages a given dimension is selected for given method, sample size, true rank and prediction criterion are shown in the Table.

Table 4.5.1 (continued)

RDA					CCA					RRR				
n = 25	MSEP		IC			MSEP		IC			MSEP		IC	
TRUE RANK = 1														
	asy	cv	asy	cv		asy	cv	asy	cv		asy	cv	asy	cv
DIM1	98	87	98	66	DIM1	87	55	90	31	DIM1	99	81	95	49
DIM2	2	11	2	26	DIM2	11	38	10	53	DIM2	1	18	5	36
DIM3	0	2	0	7	DIM3	0	6	0	15	DIM3	0	1	0	15
DIM4	0	0	0	1	DIM4	2	1	0	1	DIM4	0	0	0	0
DIM5	0	0	0	0	DIM5	0	0	0	0	DIM5	0	0	0	0
TRUE RANK = 2														
	asy	cv	asy	cv		asy	cv	asy	cv		asy	cv	asy	cv
DIM1	0	0	0	0	DIM1	0	0	0	0	DIM1	0	0	0	0
DIM2	96	85	94	51	DIM2	83	60	83	41	DIM2	94	79	93	45
DIM3	4	14	6	35	DIM3	13	35	15	51	DIM3	6	20	6	45
DIM4	0	1	0	13	DIM4	3	4	2	7	DIM4	0	1	0	10
DIM5	0	0	0	1	DIM5	1	1	0	1	DIM5	0	0	1	0
TRUE RANK = 3														
	asy	cv	asy	cv		asy	cv	asy	cv		asy	cv	asy	cv
DIM1	0	0	0	0	DIM1	0	0	0	0	DIM1	0	0	0	0
DIM2	0	0	0	0	DIM2	1	0	1	0	DIM2	0	0	0	0
DIM3	95	93	92	51	DIM3	81	72	77	65	DIM3	95	83	94	56
DIM4	5	7	7	37	DIM4	14	25	19	32	DIM4	4	16	6	39
DIM5	0	0	1	12	DIM5	4	3	3	3	DIM5	1	1	0	5
TRUE RANK = 4														
	asy	cv	asy	cv		asy	cv	asy	cv		asy	cv	asy	cv
DIM1	0	0	0	0	DIM1	0	0	0	0	DIM1	0	0	0	0
DIM2	0	0	0	0	DIM2	0	0	0	0	DIM2	0	0	0	0
DIM3	0	0	0	0	DIM3	0	0	0	0	DIM3	0	0	0	0
DIM4	91	97	67	42	DIM4	87	85	77	77	DIM4	90	90	88	76
DIM5	9	3	33	58	DIM5	13	15	23	23	DIM5	10	10	12	24
TRUE RANK = 5														
	asy	cv	asy	cv		asy	cv	asy	cv		asy	cv	asy	cv
DIM1	0	0	0	0	DIM1	0	0	0	0	DIM1	0	0	0	0
DIM2	0	0	0	0	DIM2	0	0	0	0	DIM2	0	0	0	0
DIM3	0	0	0	0	DIM3	0	0	0	0	DIM3	0	0	0	0
DIM4	0	0	0	0	DIM4	0	0	0	0	DIM4	0	0	0	0
DIM5	100	100	100	100	DIM5	100	100	100	100	DIM5	100	100	100	100

RDA					CCA					RRR				
n = 50	MSEP		IC			MSEP		IC			MSEP		IC	
TRUE RANK = 1														
	asy	cv	asy	cv		asy	cv	asy	cv		asy	cv	asy	cv
DIM1	98	83	98	74	DIM1	96	78	94	66	DIM1	97	81	97	72
DIM2	2	17	2	24	DIM2	4	20	6	30	DIM2	3	18	3	26
DIM3	0	0	0	2	DIM3	0	2	0	4	DIM3	0	1	0	2
DIM4	0	0	0	0	DIM4	0	0	0	0	DIM4	0	0	0	0
DIM5	0	0	0	0	DIM5	0	0	0	0	DIM5	0	0	0	0
TRUE RANK = 2														
	asy	cv	asy	cv		asy	cv	asy	cv		asy	cv	asy	cv
DIM1	0	0	0	0	DIM1	0	0	0	0	DIM1	0	0	0	0
DIM2	94	89	95	70	DIM2	93	80	93	74	DIM2	93	83	94	72
DIM3	6	11	5	28	DIM3	7	20	7	26	DIM3	7	17	6	26
DIM4	0	0	0	2	DIM4	0	0	0	0	DIM4	0	0	0	2
DIM5	0	0	0	0	DIM5	0	0	0	0	DIM5	0	0	0	0
TRUE RANK = 3														
	asy	cv	asy	cv		asy	cv	asy	cv		asy	cv	asy	cv
DIM1	0	0	0	0	DIM1	0	0	0	0	DIM1	0	0	0	0
DIM2	0	0	0	0	DIM2	0	0	0	0	DIM2	0	0	0	0
DIM3	92	89	90	79	DIM3	89	82	87	78	DIM3	91	83	89	77
DIM4	5	11	6	16	DIM4	7	15	9	19	DIM4	6	16	7	20
DIM5	3	0	4	5	DIM5	4	3	4	3	DIM5	3	1	4	3
TRUE RANK = 4														
	asy	cv	asy	cv		asy	cv	asy	cv		asy	cv	asy	cv
DIM1	0	0	0	0	DIM1	0	0	0	0	DIM1	0	0	0	0
DIM2	0	0	0	0	DIM2	0	0	0	0	DIM2	0	0	0	0
DIM3	0	0	0	0	DIM3	0	0	0	0	DIM3	0	0	0	0
DIM4	91	94	83	79	DIM4	90	87	91	84	DIM4	91	88	91	82
DIM5	9	6	17	21	DIM5	10	13	9	16	DIM5	9	12	9	18
TRUE RANK = 5														
	asy	cv	asy	cv		asy	cv	asy	cv		asy	cv	asy	cv
DIM1	0	0	0	0	DIM1	0	0	0	0	DIM1	0	0	0	0
DIM2	0	0	0	0	DIM2	0	0	0	0	DIM2	0	0	0	0
DIM3	0	0	0	0	DIM3	0	0	0	0	DIM3	0	0	0	0
DIM4	0	0	0	0	DIM4	0	0	0	0	DIM4	0	0	0	0
DIM5	100	100	100	100	DIM5	100	100	100	100	DIM5	100	100	100	100

Table 4.5.1 (continued)

RDA					CCA					RRR				
n = 75	MSEP		IC			MSEP		IC			MSEP		IC	
TRUE RANK = 1														
	asy	cv	asy	cv		asy	cv	asy	cv		asy	cv	asy	cv
DIM1	98	82	98	79	DIM1	96	75	94	71	DIM1	97	79	97	75
DIM2	2	17	2	16	DIM2	3	23	5	26	DIM2	3	19	3	22
DIM3	0	1	0	5	DIM3	1	2	1	3	DIM3	0	2	0	3
DIM4	0	0	0	0	DIM4	0	0	0	0	DIM4	0	0	0	0
DIM5	0	0	0	0	DIM5	0	0	0	0	DIM5	0	0	0	0
TRUE RANK = 2														
	asy	cv	asy	cv		asy	cv	asy	cv		asy	cv	asy	cv
DIM1	0	0	0	0	DIM1	0	0	0	0	DIM1	0	0	0	0
DIM2	93	85	92	67	DIM2	91	73	91	68	DIM2	92	78	93	67
DIM3	6	15	7	29	DIM3	7	25	8	30	DIM3	7	21	6	30
DIM4	1	0	1	4	DIM4	2	2	1	2	DIM4	1	1	1	3
DIM5	0	0	0	0	DIM5	0	0	0	0	DIM5	0	0	0	0
TRUE RANK = 3														
	asy	cv	asy	cv		asy	cv	asy	cv		asy	cv	asy	cv
DIM1	0	0	0	0	DIM1	0	0	0	0	DIM1	0	0	0	0
DIM2	0	0	0	0	DIM2	0	0	0	0	DIM2	0	0	0	0
DIM3	93	91	85	64	DIM3	88	85	85	79	DIM3	90	83	88	74
DIM4	5	9	10	31	DIM4	11	14	13	20	DIM4	9	15	10	23
DIM5	2	0	5	5	DIM5	1	1	2	1	DIM5	1	2	2	3
TRUE RANK = 4														
	asy	cv	asy	cv		asy	cv	asy	cv		asy	cv	asy	cv
DIM1	0	0	0	0	DIM1	0	0	0	0	DIM1	0	0	0	0
DIM2	0	0	0	0	DIM2	0	0	0	0	DIM2	0	0	0	0
DIM3	0	0	0	0	DIM3	0	0	0	0	DIM3	0	0	0	0
DIM4	88	93	72	70	DIM4	87	87	87	87	DIM4	88	86	88	80
DIM5	12	7	28	30	DIM5	13	13	13	13	DIM5	12	14	12	20
TRUE RANK = 5														
	asy	cv	asy	cv		asy	cv	asy	cv		asy	cv	asy	cv
DIM1	0	0	0	0	DIM1	0	0	0	0	DIM1	0	0	0	0
DIM2	0	0	0	0	DIM2	0	0	0	0	DIM2	0	0	0	0
DIM3	0	0	0	0	DIM3	0	0	0	0	DIM3	0	0	0	0
DIM4	0	0	0	0	DIM4	0	0	0	0	DIM4	0	0	0	0
DIM5	100	100	100	100	DIM5	100	100	100	100	DIM5	100	100	100	100

Although crossvalidation is intuitively appealing and has many other advantages, for example for model diagnostics, it does not seem to be the ideal method for model selection. Hence Efron (1983) proposed other less variable resampling estimates. On the other hand, the setup of these simulation studies might give a too optimistic impression of the asymptotic criterion, since its behavior under non–normal errors, wrong error structures and other model deviations has not been studied.

With respect to the three models (RDA, CCA, RRR) corresponding to the three error structures (4.5.1a–c), one sees that the model with the fewest free parameters (RDA) gives the most precise rank selections, the model with the most free parameters (CCA) the least precise rank selection, and the model with the intermediate number of free parameter (RRR) lies in the middle of the two.

4.5.3. Mean Squared Error of Prediction for Rank Selection

4.5.3.1. Estimation of the Mean Squared Error of Prediction

The mean squared error of prediction MSEP has been introduced in section 2.3 as another alternative measure of the predictive power of a predictor. For reduced rank regression models, the mean squared error of prediction (2.3.7) is defined as

(4.5.9) MSEP

$$:= 1/n \; E_{Y|X} E_{\tilde{Y}|X} \, \| (\tilde{Y} - 1_n \hat{\mu}(Y)' - X\hat{A}(Y)\hat{B}(Y)')\Gamma^{-1/2} \|^2,$$

where $\tilde{Y}|X$ is independent and identically distributed as $Y|X$.

The fit, asymptotic and crossvalidation estimates introduced in section 2.3.5.1 are obtained as

(4.5.10) $\mathrm{MSEP}_{\mathrm{FIT}}$

$$:= 1/n \; \| (Y - 1_n \hat{\mu}(Y)' - X\hat{A}(Y)\hat{B}(Y)')\Gamma^{-1/2} \|^2 = \mathrm{tr}[\hat{\Sigma}(Y)\Gamma^{-1}] \; .$$

(4.5.11) $$\mathrm{MSEP}_{\mathrm{ASY}} := \mathrm{MSEP}_{\mathrm{FIT}} - \hat{\delta} \, .$$

(4.5.12) $$\mathrm{MSEP}_{\mathrm{CV}} := 1/n \; \mathrm{tr}[\, (Y - \hat{\tilde{Y}}_{(.)})'(Y - \hat{\tilde{Y}}_{(.)}) \, \Gamma^{-1} \,] \, .$$

The asymptotic correction $\hat{\delta}$ in (4.5.11) is an estimate of the covariance between the observed and the predicted response matrix

$$\delta := -2 \operatorname{tr}(\operatorname{cov}(Y, \hat{Y})\Gamma^{-1}).$$

Using the notation of section 2.3.5.1 with

$$Y = Y_0 + E,$$

the Taylor expansion of $\hat{M}(Y)$ is given by

$$\operatorname{vec}(\hat{M}(Y)) \approx \operatorname{vec}(\hat{M}(Y_0)) + C(Y_0)\operatorname{vec}(E).$$

Since the estimated coefficient matrix depends on E only through $X'E$, one knows that

$$C(Y_0) = \tilde{C}(Y_0)(I_q \otimes X').$$

Hence its covariance is approximately given by

$$\operatorname{cov}(\hat{M}(Y)) \approx \tilde{C}(Y_0)(\Sigma \otimes (X'X))\tilde{C}(Y_0)'.$$

A comparison with the asymptotic covariance matrix derived in section 4.3, using (4.3.31–32) shows that approximately,

(4.5.13) $C(Y_0) \approx G^{+}G(I_q \otimes (X'X)^{1/2}RX')$, where
$G := ((B \otimes I_p) \; (I_q \otimes A))$ and
$R(\Sigma \otimes (X'X))R' = \Sigma \otimes (X'X)$ for some unknown matrix R.

Using (2.3.14) and replacing the unknown parameters in (4.5.14) by their maximum likelihood estimates, one then obtains

(4.5.14) $\mathrm{MSEP}_{ASY} := \mathrm{MSEP}_{FIT} + 2 \cdot \operatorname{tr}((\hat{\Sigma}\Gamma^{-1} \otimes X)\hat{C}(Y_0))$
$= \mathrm{MSEP}_{FIT} + 2 \cdot \operatorname{tr}((\hat{\Sigma}\Gamma^{-1} \otimes I_p)\hat{G}^{+}\hat{G})$,
independent of the matrix R.

Note that if one would choose $\Gamma^{-1} = \hat{\Sigma}$, then the second part of (4.5.15) would reduce to

$$2(p \cdot r + q \cdot r - r^2).$$

This is exactly twice the numer of free parameters of the coefficient matrix M (compare with (4.5.7)).

4.5.3.2. A Simulation Study

In section 4.5.2, the information criterion IC has been proposed for the rank selection of the reduced rank regression model. Alternatively, another prediction criterion, the mean squared error of prediction MSEP, can be used for rank selection. Again, the fit estimate is useless for selection purposes and only the asymptotic and crossvalidation estimates will be considered (see section 2.3.5.1). Since IC is invariant under arbitrary affine transformation of the responses, the error covariance matrix in (4.3.7) could be set proportional to the identity matrix, without loss of generality. Unfortunately the MSEP is only invariant to shifting. Hence using the same simulation setup as for the information criterion in section 4.5.2.2, somewhat less general results will be obtained. The weights for the MSEP defined by (2.3.2) are used in the following.

To judge the goodness of the asymptotic and crossvalidation estimates as selection procedures, their ability to select the correct rank is investigated. In the part of Table 4.5.1 referring to the MSEP, the proportion of correct selections for given true rank is shown for the different parameter settings. The conclusions are very similar to those obtained for the information criterion (see section 4.5.2.2). Comparing MSEP and IC, one sees that the crossvalidation estimate based on the MSEP seems to result in a much more precise selection of the correct rank than the crossvalidation estimate based on the IC. An explanation for this better behavior of the MSEP may be that for the estimation of IC, the error covariance matrix has to be estimated as well, resulting in higher variability.

4.5.4. Example

For the illustrative insecticide example from the QSAR area presented in section 1.6, a reduced rank regression model (4.3.7) of rank 1 with arbitrary error covariance matrix (canonical correlation analysis, CCA, see section 4.3.8.3) seems to be adequate. The rank selection procedures can be used to check if this a–priori knowledge is compatible with the data, i.e. if this model has at least the same predictive power as a full rank model. The estimates of the information criterion IC as well as of the mean squared error of prediction MSEP are shown in Table 4.5.2.

Table 4.5.2 Estimated prediction criteria for reduced rank regression(CCA)

MSEP is mean R^2 (2.3.2) (maximal values in bold face)

	DIM1	DIM2		DIM1	DIM2
MSEP	0.728	0.767	IC	-3.73	-4.34
MSEPasy	**0.519**	0.515	ICasy	**-2.77**	-2.78
MSEPcv	0.357	**0.368**	ICcv	-2.47	**-2.02**

The results are not clear–cut, since the asymptotic estimates favour a reduced rank regression model of rank 1, while the crossvalidation estimates indicate that a full rank model gives slightly better predictions. However, since the predictive power is very similar for both models, the smaller model is chosen.

As discussed in section 2.6, if the selection of the rank is not part of the method, a too optimistic assessment of the predictive power is typically obtained. Hence reduced rank regression with automatic rank selection by either the IC or the MSEP criterion is used, where the criteria are either estimated by asymptotic approximation or crossvalidation. The methods with automatic rank selection are then assessed using crossvalidation. In Table 4.5.3 one sees again, that asymptotic estimates tend to select the smaller model while the model of full rank is selected by the crossvalidation estimates.

Table 4.5.3 Estimated prediction criteria for reduced rank regression (CCA)
MSEP is mean R^2 (2.3.2)

CCA with automatic dimension selection			
by MSEPasy :	MSEPcv 0.374		
	selected dimensions	1	2
	number of selections	23	4
by MSEPcv :	MSEPcv 0.381		
	selected dimensions	1	2
	number of selections	6	21
by ICasy :	ICcv -2.89		
	selected dimensions	1	2
	number of selections	13	14
by ICcv :	ICcv -2.15		
	selected dimensions	1	2
	number of selections	1	26

Figure 4.5.1 *Crossvalidation predictions for the 1– and 2–dimensional reduced rank regression models. Ideally, the points lie on a diagonal.*

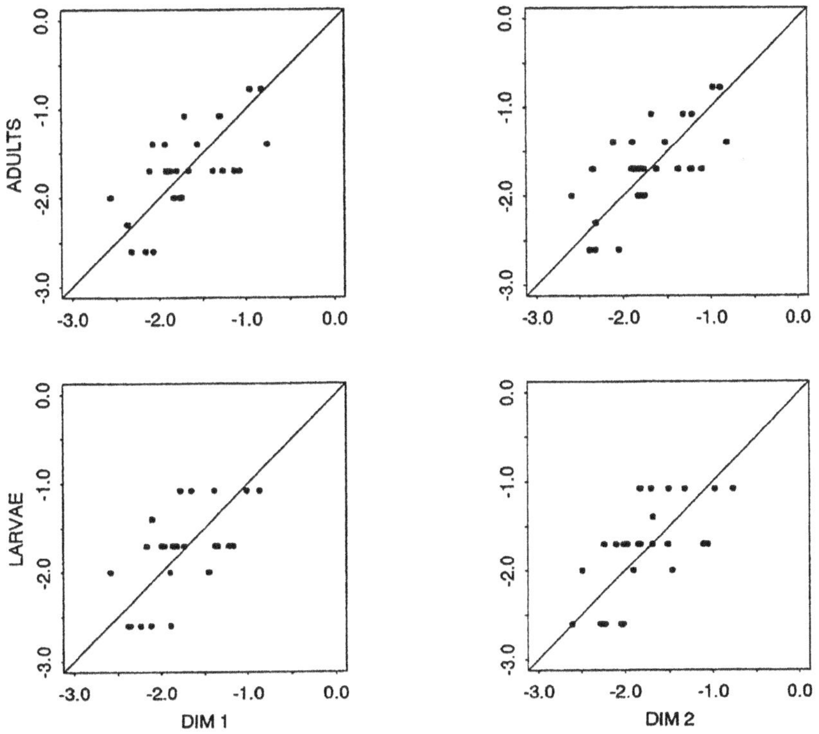

The predictive power of the reduced rank regression model seems to be somewhat smaller than the predictive power of the heuristic multivariate prediction methods (see chapter 3) principal component regression PCR and partial least squares PLS, as can be seen from a comparison with Tables 3.5.1 and 3.5.2.

One reason for this may be the colinearity of the predictor variables which can be better handled by the heuristic methods PCR and PLS. In the next section, variable selection will be used in combination with reduced rank regression models, which then seems to result in a similar predictive power as PCR and PLS.

The prediction models corresponding to the 1– and 2–dimensional solutions, respectively, are graphically compared in Figures 4.5.1 and

4.5.2 (see section 3.5). For most observations, the differences are relatively small, especially for the ADULTS. However for the LARVAE, at least 3 observations are very differently predicted by the two models (mainly vertical deviations in Figure 4.5.2).

Figure 4.5.2 *Crossvalidation residuals for the 1– and 2–dimensional reduced rank regression models. Static interpolation of corresponding observations (1 DIM :• ; 2 DIM : •). For similar methods, the line segments are small.*

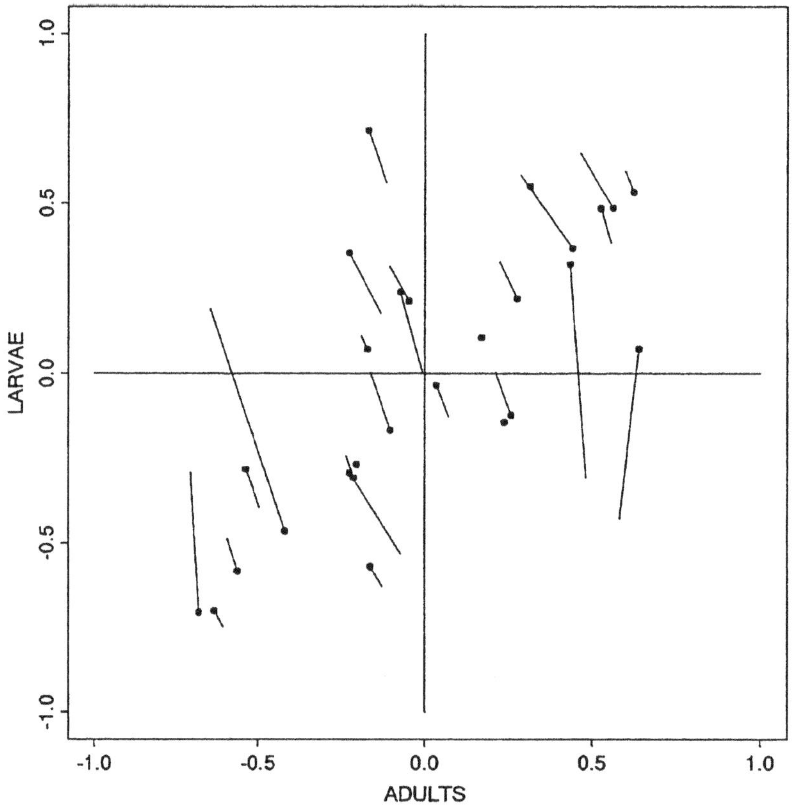

4.6. Variable Selection for Reduced Rank Regression Models

4.6.1. Prediction Criteria for Variable Selection

In section 2.5.3, many reasons for the use of variable selection in general have been given which are all relevant for reduced rank regression models. A forward variable selection procedure will be considered in the following that results in an ordering of the predictor variables according to their importance as predictors. Their importance is judged by the prediction criteria introduced in section 2.3, namley the information criterion IC and the mean squared error of prediction MSEP. The variable selection procedure consists then of the selection of that number of variables which corresponds to the prediction model with smallest estimated loss (see section 2.3.5). More precisely the following forward variable selection procedure is used :

> First, that predictor variable is selected which gives the minimal estimated expected loss when used as a single predictor. Then, the variable is added which gives minimal estimated expected loss when used as a predictor together with the previously selected variable, and so on. The variable selection ends, if all variables have been selected or after a prespecified number of variables has been selected.

Note that the procedure uses internally an estimate of the expected loss to order the variables according to their importance as predictors. Since the computational cost prohibits the use of a crossvalidation estimate of the expected loss, the fit or the asymptotic estimate can be used (see section 2.3.5). It is easy to see that both procedures lead to exactly the same ordering of the variables so that the fit estimate will be used in the following. To define a specific variable selection procedure, the used internal estimate of the expected loss has to be specified which can either be based on the information criterion IC (2.3.8) or the mean squared error of prediction MSEP (2.3.7).

The forward variable selection results only in an ordering of the variables. A prediction model is specified once the number of used variables is fixed to a specific value. Similar to the rank selection in section 4.5, each selected number of variables corresponds to a different prediction model and a prediction criterion, an estimate of the MSEP or the

IC, can be used for the selection. For consistency reasons, this external prediction criterion for the selection of the number of variables should be the same as the one used internally for the ordering of the variables, that is either the MSEP or the IC criterion.

4.6.2. A Simulation Study

If there are p predictor variables, the estimated expected loss associated with the prediction model corresponding to k selected variables is denoted by IC(k) and MSEP(k), $k=1, \dots, p$, respectively, depending on whether the information criterion IC or the mean squared error of prediction MSEP is used. That model is then selected for which the estimated MSEP or IC is minimal. The fit estimate should not be used for the selection of the number of variables, since then always the maximal number of variables is selected.

A small simulation study is intended to give some insight into the goodness of the asymptotic and crossvalidation estimates as variable selectors. Since the results for the three different error structures investigated in section 4.5 were relatively similar, only the case of an error covariance proportional to the identity matrix is considered (redundancy analysis, RDA, see section 4.3.8.2).

For the simulation of the reduced rank regression model (4.3.7), the parameters μ, Σ are set to $\mu = 0$ and $\Sigma = I_q$, the number of predictor variables is fixed to $p = 10$, and the number of response variables to $q=5$. The rank is set to $r=1$ and the true number of important variables and the sample size are varied as follows.

- true number of important variables $r = 1, 2, 3, 4, 5$;
- sample size $n = 25, 50, 75$.

For fixed true number of important variables and sample size, the predictor matrix and the parameters are set as follows.

- $n \times p$ predictor matrix $X := (\tilde{X} - 1/n \; 1_n 1_n' \tilde{X})$, where $\tilde{X} \sim N(0, I_{np})$;
- p–dimensional parameter vector $A_i := 1$ if $i < (r+1)$ and $A_i := 0$ otherwise;
- q–dimensional parameter vector $B := f \; \tilde{B}$, where $\tilde{B} := 1_q$ and $f := \sqrt{1/((1 - R^2) \, \| X A \tilde{B}' \|^2 / nq)}$ with $R^2 := 0.8$.

Table 4.6.1 Simulation

For given sample size (n=25,50,75) and true number of important variables (1-5), the number of important variables is estimated by four prediction criteria :

MSEPasy : asymptotic estimate of the mean squared error of prediction,

MSEPcv : crossvalidation estimate of the mean squared error of prediction,

ICasy : asymptotic estimate of the information criterion,

ICcv : crossvalidation estimate of the information criterion.

The percentages a number of variables is selected shown below. The ordering of the ten variables (x.1,...,x.10) is the one obtained by using the fit criterion.

	n = 25					n = 50					n = 75			
TRUE NUMBER OF IMPORTANT VARIABLES = 1														
	MSEP		IC			MSEP		IC			MSEP		IC	
	asy	cv	asy	cv		asy	cv	asy	cv		asy	cv	asy	cv
x.1	65	22	61	20	x.1	56	24	56	22	x.1	53	25	53	24
x.5	22	37	23	38	x.2	20	33	20	33	x.3	20	37	21	33
x.2	7	27	8	19	x.4	8	20	8	20	x.4	18	21	17	21
x.9	3	9	3	13	x.5	8	14	8	14	x.9	2	11	2	17
x.6	2	4	4	7	x.3	2	4	2	5	x.8	3	4	3	3
x.8	1	0	1	3	x.9	2	4	2	5	x.7	3	1	3	1
x.3	0	1	0	0	x.8	3	1	3	1	x.10	1	1	1	1
x.4	0	0	0	0	x.10	1	0	1	0	x.2	0	0	0	0
x.10	0	0	0	0	x.7	0	0	0	0	x.5	0	0	0	0
x.7	0	0	0	0	x.6	0	0	0	0	x.6	0	0	0	0

Table 4.6.1 (continued)

	n = 25					n = 50					n = 75			
TRUE NUMBER OF IMPORTANT VARIABLES = 2														
	MSEP		IC			MSEP		IC			MSEP		IC	
	asy	cv	asy	cv		asy	cv	asy	cv		asy	cv	asy	cv
x.1	0	0	0	0	x.1	0	0	0	0	x.1	0	0	0	0
x.2	64	28	62	27	x.2	61	31	58	29	x.2	49	26	48	25
x.5	26	37	28	36	x.4	14	26	15	25	x.3	26	42	25	39
x.9	5	20	5	17	x.5	9	26	10	22	x.4	16	20	18	22
x.6	2	8	2	10	x.3	7	10	8	15	x.9	4	7	4	8
x.4	2	7	2	10	x.9	3	5	3	5	x.8	1	4	1	5
x.10	1	0	1	0	x.8	4	2	3	4	x.7	4	1	4	1
x.3	0	0	0	0	x.10	2	0	2	0	x.10	0	0	0	0
x.7	0	0	0	0	x.7	0	0	1	0	x.5	0	0	0	0
x.8	0	0	0	0	x.6	0	0	0	0	x.6	0	0	0	0
TRUE NUMBER OF IMPORTANT VARIABLES = 3														
	MSEP		IC			MSEP		IC			MSEP		IC	
	asy	cv	asy	cv		asy	cv	asy	cv		asy	cv	asy	cv
x.1	0	0	0	0	x.3	0	0	0	0	x.2	0	0	0	0
x.3	0	0	0	0	x.1	0	0	0	0	x.1	0	0	0	0
x.2	66	30	68	29	x.2	62	37	62	35	x.3	50	32	50	31
x.5	25	39	21	37	x.4	15	23	15	25	x.9	28	39	29	39
x.9	4	22	7	19	x.5	6	25	7	21	x.8	15	20	14	19
x.6	0	7	0	13	x.8	6	8	6	12	x.4	3	7	3	8
x.4	3	2	3	1	x.9	6	7	6	6	x.7	2	2	2	3
x.10	1	0	1	0	x.10	2	0	2	1	x.10	2	0	2	0
x.8	1	0	0	1	x.7	2	0	2	0	x.5	0	0	0	0
x.7	0	0	0	0	x.6	1	0	0	0	x.6	0	0	0	0

	n = 25					n = 50					n = 75			
TRUE NUMBER OF IMPORTANT VARIABLES = 4														
	MSEP		IC			MSEP		IC			MSEP		IC	
	asy	cv	asy	cv		asy	cv	asy	cv		asy	cv	asy	cv
x.1	0	0	0	0	x.3	0	0	0	0	x.4	0	0	0	0
x.4	0	0	0	0	x.2	0	0	0	0	x.2	0	0	0	0
x.3	0	0	0	0	x.4	0	0	0	0	x.1	0	0	0	0
x.2	60	37	59	32	x.1	65	40	65	38	x.3	50	36	50	34
x.5	21	37	23	35	x.5	15	27	15	27	x.9	29	35	30	37
x.9	11	23	12	29	x.8	9	20	9	22	x.8	14	23	14	21
x.6	7	3	4	3	x.9	7	10	7	10	x.7	3	5	2	6
x.10	1	0	2	0	x.10	3	3	3	3	x.10	4	1	4	2
x.8	0	0	0	1	x.7	1	0	1	0	x.5	0	0	0	0
x.7	0	0	0	0	x.6	0	0	0	0	x.6	0	0	0	0
TRUE NUMBER OF IMPORTANT VARIABLES = 5														
	MSEP		IC			MSEP		IC			MSEP		IC	
	asy	cv	asy	cv		asy	cv	asy	cv		asy	cv	asy	cv
x.5	0	0	0	0	x.3	0	0	0	0	x.2	0	0	0	0
x.1	0	0	0	0	x.1	0	0	0	0	x.1	0	0	0	0
x.4	0	0	0	0	x.4	0	0	0	0	x.5	0	0	0	0
x.2	0	0	0	0	x.5	0	0	0	0	x.4	0	0	0	0
x.3	50	44	51	40	x.2	67	47	66	44	x.3	55	42	58	46
x.9	29	36	29	39	x.9	19	30	20	32	x.9	27	38	25	33
x.6	15	18	13	18	x.8	9	18	8	19	x.8	14	18	13	16
x.10	4	2	5	2	x.10	5	5	5	5	x.7	3	2	3	5
x.8	1	0	1	1	x.7	0	0	1	0	x.10	1	0	1	0
x.7	1	0	1	0	x.6	0	0	0	0	x.6	0	0	0	0

For each fixed setting of the above parameters, the simulation proceeds as follows. Let N be a large integer, N=100, say, then

for i=1, 2, ... , N,

sample $Y^{(i)} \sim N(XAB', I_{nq})$,
for assumed number of variables j=1, 2, ... , p , calculate
$IC^{(i)}_{ASY}(j)$ and $IC^{(i)}_{CV}(j)$ as in (4.5.5), (4.5.8) and
$MSEP^{(i)}_{ASY}(j)$ and $MSEP^{(i)}_{CV}(j)$ as in (4.5.11), (4.5.12).

To judge the goodness of the four selection procedures, their ability to select the correct number of variables is investigated. In Table 4.6.1 the proportions of correct selections for given true number of important variables is shown for the different parameter settings. One sees that too few variables are never selected by the criteria, however often too many variables are chosen. Again, asymptotic criteria are usually more precise than crossvalidation criteria in the selection of the correct number of variables (compare the discussion in section 4.5.2.2). The differences between the criteria based on the mean squared error of prediction MSEP and the information criterion IC, respectively, are small.

4.6.3. Example

Again the QSAR example discussed in section 1.6 is used to illustrate the forward variable selection procedure. Since a reduced rank regression model of rank 1 with arbitrary error covariance matrix (canonical correlation analysis, CCA, see section 4.3.8.3) seems to be acceptable, variable selection with this fixed setting is used. The estimates of the information criterion IC as well as of the mean squared error of prediction MSEP are shown in the first part of Table 4.6.2. One sees that models with 4 to 6 variables seem to give the best prediction models.

As discussed in section 2.6, if the selection of the number of variables is not part of the method, a usually too optimistic assessment of the predictive power is obtained. Hence variable selection with automatic selection of the optimal number of variables by the asymptotic or crossvalidation estimates of either the IC or the MSEP criterion is used. These models with automatic variable selection are then assessed by crossvalidation. In the second part of Table 4.6.2 one sees that typically 4 to 6 variables are selected and that the predictive power of the method with automatic selection of the number of variables is in fact smaller.

Table 4.6.2 Estimated prediction criteria for reduced rank regression with variable selection (CCAv)

MSEP is mean R^2 (2.3.2) (maximal values in bold face)

MSEP	I1	I2	R	F	MR	L	IP23	B1	Pi
fit	0.130	0.450	0.603	0.670	0.690	0.714	0.722	0.726	0.728
asy	0.104	0.403	0.534	**0.581**	0.579	**0.581**	0.568	0.550	0.519
cv	-0.437	0.050	0.362	0.299	0.351	**0.422**	0.413	0.393	0.357

IC	I1	I2	R	F	MR	L	IP23	B1	Pi
fit	-2.466	-2.947	-3.304	-3.506	-3.606	-3.665	-3.681	-3.688	-3.734
asy	-2.096	-2.502	-2.785	-2.913	**-2.939**	-2.924	-2.866	-2.799	-2.771
cv	-1.664	-2.091	-2.586	-2.439	-2.533	**-2.646**	-2.589	-2.557	-2.469

MSEP is mean R^2 (2.3.2)

automatic selection of the number of variables

by MSEPasy :	MSEPcv 0.287								
	selected number of variables	4	5	6	7				
	number of selection	16	6	3	2				
by MSEPcv :	MSEPcv 0.288								
	selected number of variables	3	4	5	6	7	8	9	
	number of selection	4	0	2	11	8	1	1	
by ICasy :	ICcv -2.41								
	selected number of variables	4	5	6	7				
	number of selection	2	22	1	2				
by ICcv :	ICcv -2.34								
	selected number of variables	2	3	4	5	6	7		
	number of selection	2	3	1	5	10	6		

In Table 4.6.3 the coefficient matrix is shown. The structural parameters I1 and I2, and the electronic parameters F and R seem to be necessary for an acceptable prediction model. The steric variables MR and/or L seem to give some small additional predictive power. Note that a comparison of the asymptotic estimates of the standard deviations of the estimated coefficients with the estimated coefficients resulted in essentially the same conclusions (see section 4.3.11).

Table 4.6.3 Coefficient matrix M for the prediction model (CCAv)

	4 VARS.		5 VARS.		6 VARS.	
	ADULTS	LARVAE	ADULTS	LARVAE	ADULTS	LARVAE
CONST	-1.685	-1.751	-1.684	-1.750	-1.684	-1.750
I1	-0.282	-0.261	-0.265	-0.244	-0.257	-0.245
I2	0.293	0.272	0.289	0.266	0.295	0.281
F	-0.122	-0.113	-0.100	-0.092	-0.044	-0.042
R	-0.203	-0.188	-0.223	-0.205	-0.253	-0.241
MR	0.000	0.000	0.084	0.077	0.217	0.207
L	0.000	0.000	0.000	0.000	-0.144	-0.137

The predictions of the test dataset are shown in Table 4.6.4. They are similar for the 3 models corresponding to 4, 5 and 6 variables, respectively. Again substances IIia and IIIia are outliers (see section 3.5). For the other substances, slightly better predictions are obtained than by the reduced rank regression models (see section 4.5.3) and the heuristic multivariate prediction methods partial least squares and principal component regression (see section 3.5).

Table 4.6.4 Predictions for the test data (outlier substances in bold face)

			4 VARS.		5 VARS.		6 VARS.	
	ADULTS	LARVAE	ADULTS	LARVAE	ADULTS	LARVAE	ADULTS	LARVAE
IIca	-2.0	>-1.7	-1.658	-1.727	-1.722	-1.785	-1.608	-1.678
IIcb	>-1.7	-2.0	-2.231	-2.257	-2.262	-2.282	-2.131	-2.176
IIia	**-2.6**	**-2.6**	**-0.826**	**-0.955**	**-0.684**	**-0.829**	**-0.481**	**-0.604**
IIIia	**>-1.7**	**-2.0**	**-0.250**	**-0.422**	**-0.115**	**-0.306**	**0.097**	**-0.052**
IIIka	>-1.7	-1.7	-1.103	-1.212	-1.057	-1.173	-1.135	-1.228
IIIla	>-1.7	>-1.7	-1.000	-1.116	-1.087	-1.201	-1.121	-1.214
IIIma	>-1.7	>-1.7	-1.530	-1.608	-1.303	-1.400	-1.136	-1.228

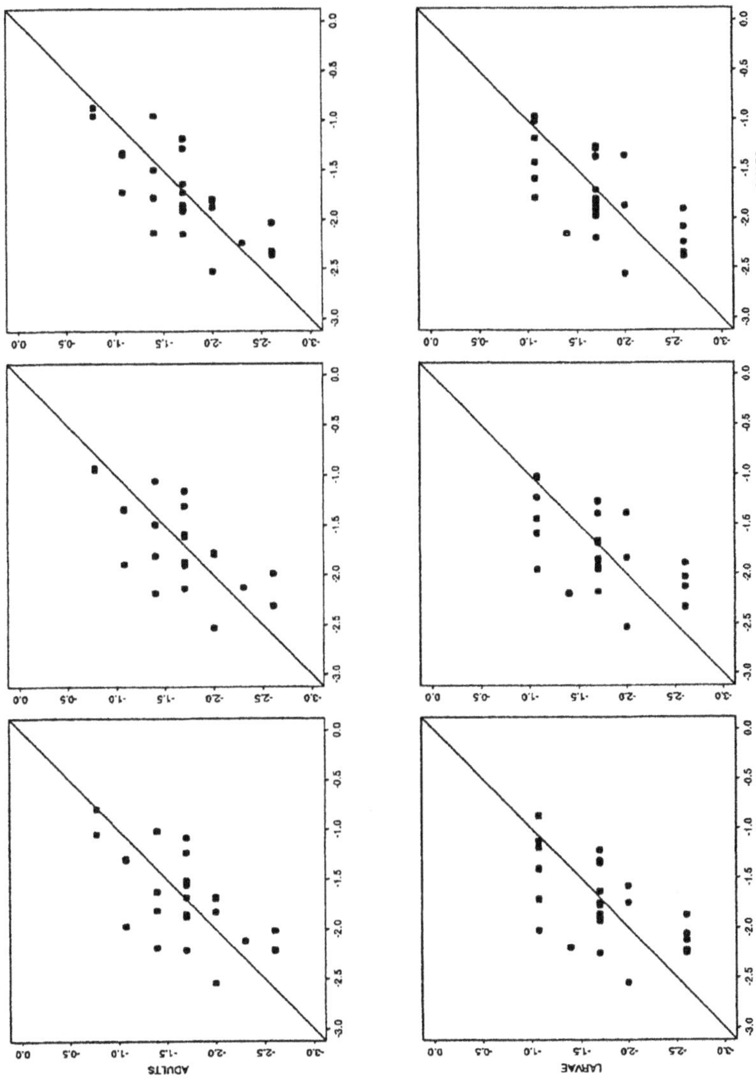

Figure 4.6.1 *Crossvalidation predictions for the 3 prediction models corresponding to 4, 5 and 6 predictor variables, respectively. Ideally, points lie on a diagonal.*

In Figures 4.6.1 and 4.6.2, the 3 prediction models are graphically compared and confirm the results obtained from the numerical measures (compare with the plots in section 3.5 and 4.5.4).

Figure 4.6.2 *Crossvalidation residuals for the 3 prediction models corresponding to 4, 5 and 6 predictor variables, respectively. Static interpolation of corresponding observations (4 VARS.:•; 5 VARS : •; 6 VARS : •).*

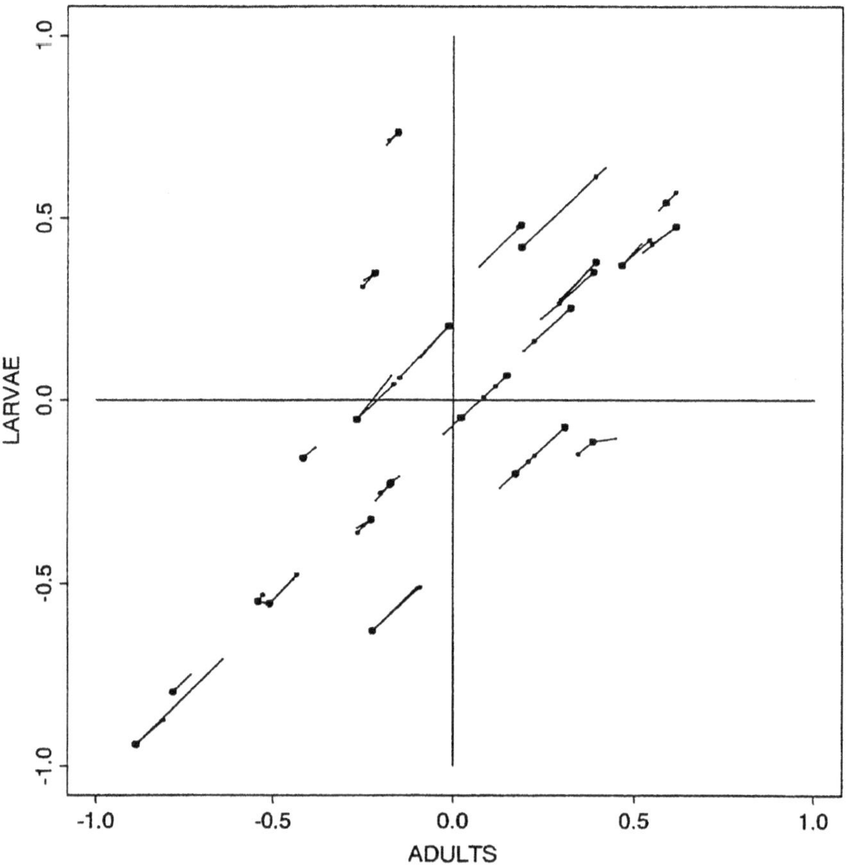

5.

Bayesian Analysis of Reduced Rank Regression

5.1. Introduction

Classical maximum likelihood analysis assumes a parametric model for the data, specified by the likelihood $p(Y|\theta)$ for parameter vector θ. The likelihood is then regarded as a function of θ and the maximum likelihood estimate $\hat{\theta} \equiv \hat{\theta}(Y)$ is the parameter value maximizing the likelihood. Unfortunately, the exact finite sample distribution of $\hat{\theta}$ is only known in simple cases and approximate distributions have to be used based on asymptotic theory. For reduced rank regression models, both maximum likelihood estimates and asymptotic distributions have been presented in the previous chapter.

The Bayesian approach assumes as well that the data can be described by the likelihood $p(Y|\theta)$, but proceedes by first defining a prior distribution $p(\theta)$ for the parameter vector θ reflecting the a–priori belief about the parameters. The posterior distribution $p(\theta|Y)$ then reflects the a–posteriori uncertainty about the parameters, that is after having seen the data. All Bayesian inference is based on this posterior distribution, which is calculated from prior and likelihood by Bayes's theorem:

$$p(\theta|Y) = \frac{p(Y|\theta)p(\theta)}{\int p(Y|\theta)p(\theta)d\theta} \propto p(Y|\theta)p(\theta)$$

A Bayesian analysis of a multivariate model with typically very many parameters is made difficult by two problems : the specification of an appropriate prior distribution for the parameters, and the integration necessary to calculate the posterior distribution.

The first difficulty is often addressed by considering non–informative priors. In the following only informative priors are used. However, these will be chosen to have small impact on the posterior distribution in relation to the information provided by the data.

The necessary multidimensional integration to obtain the posterior distribution has been for a long time the main obstacle in the practical use of Bayesian methodology, since for all but the most simple models, unhandy numerical integration procedures had to be used. However, the recent introduction of Markov chain Monte Carlo methods, especially the Gibbs sampler (Gelfand and Smith, 1990) has dramatically changed the situation. A Bayesian analysis of a realistic model is now not much more difficult to obtain than an analysis by classical statistical methods.

The following Bayesian analysis of reduced rank regression is essentially taken from Schmidli (1994). First, a Bayesian reduced rank regression model is presented in section 5.2, emphasizing the importance of an adequate parametrization. The model is specified by the full conditional posterior distributions which are needed for the Gibbs sampling procedure. Three important reduced rank regression models are considered, differing by their error covariance structure. Then the predictive distribution is introduced and prediction criteria for rank selection are proposed; variable selection methods will not be used. In section 5.3, the calculation of the posterior and predictive distribution by Gibbs sampling is presented and finally the methods are illustrated by a detailed example. For a Bayesian analysis of reduced rank regression with improper priors and a different parametrization, see Geweke (1993).

In the following, for random matrices U and V, $[U|V]$ denotes the conditional distribution of U given V, and $[U]$ denotes the marginal distribution of U. An $n \times q$ random matrix U has a matrix normal distribution denoted by $N(M, \Omega)$, if its vectorization $vec(U)$ has a multivariate normal distribution with nq–dimensional mean vector $vec(M)$ and $nq \times nq$ covariance matrix Ω, where M is an $n \times q$ matrix. The inverse Wishart distribution with degrees of freedom n and scale matrix Ω is denoted by $W^{-1}(n, n\Omega)$.

5.2. The Reduced Rank Regression Model

A Bayesian reduced rank regression model is uniquely defined by the specification of the full conditional posterior distributions of the parameters (see Besag, 1974). These full conditional distributions are needed for the application of the Gibbs sampling procedure, a Markov chain Monte Carlo method to approximately calculate the joint posterior distribution. From this joint posterior distribution, the predictive distribution can be obtained.

5.2.1. Likelihood

Suppose that an $n \times q$ response matrix Y is observed as a realization of the random $n \times q$ matrix Y. The reduced rank regression model assumes that the random matrix Y conditional on the parameters has a matrix normal distribution (see (4.3.7)), shown again for ease of reference

(5.2.1) $[Y \mid A, B, \Sigma, \mu] = N(1_n\mu' + XAB', \Sigma \otimes I_n)$, where

X is a fixed known $n \times p$ matrix of predictors of full rank with $X'1_n = 0$,
r is a known integer (rank), with $1 \leq r < \min(p, q)$,
A is a $p \times r$ matrix of unknown parameters of rank r,
B is a $q \times r$ matrix of unknown parameters of rank r,
μ is a q–dimensional vector of unknown parameters (mean),
Σ is a positive definite $q \times q$ matrix of unknown parameters (error covariance matrix),
I_n is the $n \times n$ identity matrix and
1_n is the n–dimensional vector of 1's.

The likelihood of the observed data given the parameters can be written as

(5.2.2) $[Y|A,B,\Sigma,\mu] \propto$

$$|\Sigma|^{-n/2} \exp[-1/2 \operatorname{tr}(Y - 1_n\mu' - XAB')'(Y - 1_n\mu' - XAB')\Sigma^{-1}].$$

5.2.2. Parametrization

The equivalence of the parameter pair (A,B) and (AF, BF'^{-1}) for any nonsingular $r \times r$ matrix F with respect to the likelihood has already been noted. In order to be able to define a probability distribution for these parameters, constraints have to be imposed to resolve this ambiguity. As discussed in section 4.3.2, many parametrizations are possible. From an interpretational standpoint, parametrization III (4.3.6) has numerous advantages and is favoured for the classical analysis of reduced rank regression. Hence this parametrization III will be used in the Bayesian analysis as well; however a more precise discussion is necessary in a Bayesian context.

The uniqueness of the parametrization can be achieved by requiring the parameter matrix A to lie in a subset Φ of the set Ψ of all $p \times r$ matrices of full rank, such that

- $\Psi \equiv \{ AF ;\ A \in \Phi,\ F$ nonsingular $r \times r$ matrix $\}$, and
- for all $A, \tilde{A} \in \Phi$: If $AF = \tilde{A}$ for some nonsingular $r \times r$ matrix F, then $A = \tilde{A}$ and F is the identity matrix.

A convenient parameter set Φ is defined as follows. Let W be a symmetric positive definite $p \times p$ matrix and let A_0 be a fixed matrix from the set

$\Phi_0 := \{A \in \Psi; A'WA = I_r\}$ with $A_0O \neq A_0$, for all orthogonal $r \times r$ matrices O different from the identity matrix. For example, A_0 can be set to the adequately scaled maximum likelihood estimate of the reduced rank regression model (5.2.1); the appropriate choice of W will be discussed in the next section, where W will be set proportional to X'X (compare with (4.3.6)).

Then the following set has the required properties :

$$\Phi := \{ A\,O(A)\,;\ A \in \Phi_0 \}, \text{ with } O(A) := \arg \min_{O \text{ orth.}} \| (A\,O - A_0) W^{1/2} \|$$

is a modified Procrustes transformation (see Sibson, 1978); compare the discussion in section 4.3.2, where an analytical solution of O(A) has been given (replacing X'X by W). The set Φ is easy to visualize geometrically for rank 1 : it corresponds to the surface of a half–ellipsoid in p–dimensional space.

Note that a mapping ϕ from the full parameter space Ψ to the constrained parameter space Φ is implicitly defined by the above requirements. For $A \in \Psi$, let the singular value decomposition of A'WA be given by $A'WA = QDQ'$ with orthogonal matrix Q and

diagonal matrix D. Defining $F(A) := Q\ D^{-1/2}$, A can be mapped first from Ψ to Φ_0 and then from Φ_0 to Φ as follows:
$A \mapsto \tilde{A} := A\ F(A) \in \Phi_0$, and $\tilde{A} \mapsto \tilde{A}\ O(\tilde{A}) \in \Phi$. The modified Procrustes transformation $O(\tilde{A})$ can be obtained as follows. Let the singular value decomposition of $A_0'W\tilde{A}$ be given by
$A_0'W\tilde{A} = U\ S\ V'$ with column orthogonal matrix U, orthogonal matrix V and diagonal matrix S. Then $O(\tilde{A}) := V$ is the required Procrustes transformation.

An alternative way to resolve the non–uniqueness of the parametrization is to fix part of the matrix A or B as discussed in section 4.3.3 (parametrization I (4.3.4) and parametrization II (4.3.5)). These parametrizations are used in Geweke (1993) for his Bayesian analysis of reduced rank regression. However, as discussed in section 4.3.3, numerical problems can be the consequence of such a parametrization which are particularly annoying if Monte Carlo methods are used to approximate the posterior distribution. Note that a mapping from the full parameter space to the constrained parameter space is implicitly defined by these parametrizations, too.

Another alternative proposed by Geweke (1993) is not to resolve the ambiguity of the parametrization, that is not to constrain the parameters at all. It is at least unclear, whether a sensible analysis can be obtained without constraints. Certainly, the interpretation of the parameters A and B is not possible any more.

5.2.3. Full Conditional Priors

Since for given A, model (5.2.1) is a standard multivariate regression model, the full conditional conjugate priors for B, Σ, μ are matrix normal and inverse Wishart distributions, respectively (see, for example, Schervish, 1989). However, since (XA) is the new predictor matrix for given A, in the context of reduced rank regression models the covariance matrix for the prior distribution of B depends on A. The full conditional priors are given by

(5.2.3) $[B' \mid A, \Sigma, \mu] = N(B^{(0)\prime}, \Sigma \otimes (A'W^{(0)}A)^{-1})$, where
$B^{(0)\prime}$ is a known $q \times r$ matrix,
$W^{(0)}$ is a known symmetric positive definite $p \times p$ matrix.

(5.2.4) $[\Sigma \mid A, B, \mu] = W^{-1}(c^{(0)}, c^{(0)}\Sigma^{(0)})$, where
$c^{(0)}$ is a known positive integer,
$\Sigma^{(0)}$ is a known symmetric positive definite $q \times q$ matrix.

(5.2.5) $[\mu \mid A, B, \Sigma] = N(\mu^{(0)}, 1/\lambda^{(0)}\Sigma)$, where
$\lambda^{(0)}$ is a known positive integer,
$\mu^{(0)}$ is a known q–dimensional vector.

For given B, model (5.2.1) has the form of the classical multivariate growth curve model introduced by Potthoff and Roy (1964). Disregarding the constraints, the conditional conjugate prior for A is matrix normal. Hence the full conditional conjugate prior for A is given by the constrained normal distribution

(5.2.6) $[A \mid B, \Sigma, \mu] = N_{\Phi}(A^{(0)}, (B'\Sigma^{-1}B)^{-1} \otimes {W^{(0)}}^{-1})$, where
$A^{(0)}$ is a known $p \times r$ matrix with $A^{(0)} \in \Phi$,
$W^{(0)}$ is the known positive definite $p \times p$ matrix specified in (5.2.3).

Here $N_{\Phi}(M, \Omega)$ denotes a matrix normal distribution constrained to the surface Φ introduced in section 5.2.2. More precisely, the density on the constraint set is chosen to be proportional to the normal density. For example, if $\Phi \subset R^n$ is the unit sphere, then $N_{\Phi}(\mu, \sigma^2 I_n)$, $\mu'\mu = 1$, is the von Mises–Fisher distribution and $N_{\Phi}(0, \Omega)$ is the Bingham distribution (see, for example, Mardia et al., 1979, ch.15).

Note that it is not necessary but consistent to use the same matrix $W^{(0)}$ in the specification of both the priors of A and B.

One problem with the attempt to define the joint distribution of the priors by specifying only the full conditional distributions of the priors is that it is generally not obvious, if the conditional distributions define in fact a joint distribution (see Besag, 1974). As noted by A.F.M.Smith (personal communication), this can be proven for the above setting by using a special constraint, namely by setting $W := \delta^{-2} W^{(0)}$ for some positive value δ^2 (see previous section 5.2.2), i.e. by requiring that $A'(\delta^{-2}W^{(0)})A = \delta^{-2}A'W^{(0)}A = I_r$. To see this, note that one then has

$$[B' \mid A, \Sigma, \mu] = N(B^{(0)\prime}, \Sigma \otimes (A'W^{(0)}A)^{-1}) = N(B^{(0)\prime}, \Sigma \otimes \delta^{-2}I_r)$$
$$= [B' \mid \Sigma],$$
$$[\Sigma \mid A, B, \mu] = [\Sigma] \text{ and}$$
$$[\mu \mid A, B, \Sigma] = [\mu \mid \Sigma].$$

Hence,

$[A, B, \Sigma, \mu] = [A \mid B, \Sigma, \mu]\,[B' \mid \Sigma]\,[\mu \mid \Sigma]\,[\Sigma]$

and the full conditional distributions define in fact a joint distribution.

Note that for a very small value of δ^2, (5.2.6) specifies approximately a uniform distribution on the constrained space, and hence reflects a–priori ignorance about the parameter A.

Since typically, the matrix $W^{(0)}$ is set to $\delta^2 \cdot X'X$ (Zellner's g–prior) or to $\delta^2 \cdot I_p$ for some small value δ^2, the constraints can be written in these two cases as $A'X'XA = I_r$ and $A'A = I_r$, respectively. The former setting is the constraint generally used in classical reduced rank regression (parametrization III (4.3.6)), while the later could be used to motivate a reduced rank ridge regression model, in a similar way as ridge regression is motivated in Lindley and Smith (1972). In the following it will always be assumed that the matrix $W^{(0)}$ is proportional to $X'X$ (Zellner's g–prior).

5.2.4. Full Conditional Posteriors

The full conditional posteriors for B, Σ, μ are derived in a similar way as for the standard multivariate regression model. They are given by

(5.2.7) $[B' \mid A, \Sigma, \mu, Y] = N(B^{(1)\prime}, \Sigma \otimes (A'W^{(1)}A)^{-1})$, where
$W^{(1)} := W^{(0)} + X'X,$
$B^{(1)\prime} := (A'W^{(1)}A)^{-1}((A'W^{(0)}A)B^{(0)\prime} + (XA)'(XA)\hat{B}'),$
with
$\hat{B}' := [(XA)'(XA)]^{-1}(XA)'(Y - 1_n\mu').$

(5.2.8) $[\Sigma \mid A, B, \mu, Y] = W^{-1}(c^{(1)}, c^{(1)}\Sigma^{(1)})$, where
$c^{(1)} := n + c^{(0)},$
$\Sigma^{(1)} :=$
$((Y - 1_n\mu' - XAB')'(Y - 1_n\mu' - XAB') + c^{(0)}\Sigma^{(0)})/c^{(1)}.$

(5.2.9) $[\mu \mid A, B, \Sigma, Y] = N(\mu^{(1)}, 1/\lambda^{(1)}\Sigma)$, where
$\lambda^{(1)} := \lambda^{(0)} + n,$
$\mu^{(1)} := 1/\lambda^{(1)}(\lambda^{(0)}\mu^{(0)} + Y'1_n).$

Note that since the posterior distribution of Σ is conditioned on B, the updating formula is different from the unconditional case (see, for example, Schervish, 1989).

The conditional posterior distribution of A can be demonstrated as follows. First, by multiplying the responses by $\Sigma^{-1/2}$, a reduced rank regression model with an error covariance equal to the identity matrix is obtained. Let $C := (C_1 C_2)$ be a nonsingular $q \times q$ matrix composed of the $q \times r$ matrix C_1 and the $q \times (q-r)$ matrix C_2 such that

$$B'\Sigma^{-1/2}C_1 = I_r \quad \text{and} \quad B'\Sigma^{-1/2}C_2 = 0_r .$$

Note that one can choose $C_1 := \Sigma^{-1/2}B(B'\Sigma^{-1}B)^{-1}$.

Then $\tilde{Y}_1 := Y\Sigma^{-1/2}C_1$
$\sim N(1_n\mu'\Sigma^{-1/2}C_1 + XA, (B'\Sigma^{-1}B)^{-1} \otimes I_n)$ and
$\tilde{Y}_2 := Y\Sigma^{-1/2}C_2 \sim N(1_n\mu'\Sigma^{-1/2}C_2, (C_2'\Sigma C_2) \otimes I_n)$ is independent of A and $\tilde{Y}_1$.

Since using the unconstrained prior and constraining the posterior distribution is equivalent to using the constrained prior to calculate the posterior distribution, the conditional posterior of A is given by the constrained normal distribution

(5.2.10) $[A \mid B, \Sigma, \mu, Y] = N_\Phi(A^{(1)}, (B'\Sigma^{-1}B)^{-1} \otimes {W^{(1)}}^{-1})$,
where
$W^{(1)} := W^{(0)} + X'X$ (as seen in (5.2.7)),
$A^{(1)} := {W^{(1)}}^{-1}(W^{(0)}A^{(0)} + X'X\hat{A})$, with
$\hat{A} := [X'X]^{-1}X'(Y - 1_n\mu')\Sigma^{-1/2}C_1$
$= [X'X]^{-1}X'(Y - 1_n\mu')\Sigma^{-1}B(B'\Sigma^{-1}B)^{-1}$.

5.2.5. Structured Error Covariance Matrix

In the above derivations of the full conditional priors and posteriors, it has been assumed that the error covariance matrix is an unstructured positive definite $q \times q$ matrix of unknown parameters. Based on additional information on the data generation procedure, some alternative structures of the error covariance matrix can be considered. Two cases are of special interest, namely the case of an error covariance matrix proportional to the identity matrix and the case of a diagonal error covariance matrix (see section 4.3.9).

a) $\Sigma = \sigma^2 I$:

The conditional prior distribution of Σ can be written as

(5.2.4*) $[\Sigma \mid A, B, \mu] = W^{-1}(c^{(0)}, c^{(0)}\sigma^{(0)}I_q)$, where
$c^{(0)}$ is a known positive integer,
$\sigma^{(0)}$ is a known positive number.

The conditional posterior distribution of Σ is then given by

(5.2.8*) $[\Sigma \mid A, B, \mu, Y] = W^{-1}(c^{(1)}, c^{(1)}\sigma^{(1)}I_q)$, where
$c^{(1)} := n + c^{(0)}$,
$\sigma^{(1)} :=$
$(1/q \; tr(Y - 1_n\mu' - XAB')'(Y - 1_n\mu' - XAB') + c^{(0)}\sigma^{(0)})/c^{(1)}$.

b) $\Sigma = Diag(\sigma_1^2, ..., \sigma_q^2)$:

The conditional prior distribution of Σ may be written as

(5.2.4**) $[\Sigma \mid A, B, \mu] = W^{-1}(c^{(0)}, c^{(0)}Diag(\sigma_1^{(0)}, ..., \sigma_q^{(0)}))$, where
$c^{(0)}$ is a known positive integer,
$\sigma_1^2, ..., \sigma_q^2$ are known positive constants.

The conditional posterior distribution of Σ is then given by

(5.2.8**) $[\Sigma \mid A, B, \mu, Y] = W^{-1}(c^{(1)}, c^{(1)}Diag(\sigma_1^{(1)}, ..., \sigma_q^{(1)}))$,
where
$c^{(1)} := n + c^{(0)}$, and for i = 1,..,q
$\sigma_i^{(1)} :=$
$([(Y - 1_n\mu' - XAB')'(Y - 1_n\mu' - XAB')]_{ii} + c^{(0)}\sigma_i^{(0)}))/c^{(1)}$.

5.2.6. Predictive Distribution

Since for QSAR applications the interest is centered on prediction of unobserved responses for given predictors, the predictive distribution is of greater interest than the posterior distribution of the parameters. For a discussion of similar issues, compare with Gelfand et al. (1992).

For fixed p–dimensional predictor vector x , the distribution of the q–dimensional random response vector y given the parameters A, B, Σ, μ is a multivariate normal distribution, i.e.

(5.2.11) $[y \mid x \,;\, A, B, \Sigma, \mu] = N(\mu' + x'AB', \Sigma)$.

Let the joint posterior distribution of the parameters be given by $[A, B, \Sigma, \mu \mid Y]$. Then the predictive distribution of y given the data Y is

(5.2.12) $[y \mid x; Y] = \int [y \mid x; A, B, \Sigma, \mu]\, [A, B, \Sigma, \mu \mid Y]$.

The predictive distribution can be used for all types of diagnostic checks of the goodness of the prediction model by comparison with the observed responses, for example by graphical methods. As an overall criterion for the goodness of prediction, several measures have been suggested in the literature, two of them are the mean squared error of prediction MSEP and the information criterion IC discussed in section 2.3. In a Bayesian context, they are defined as follows.

(5.2.13) MSEP :=

$$1/n \sum_{i=1}^{n} \| (E_{y|x_i,Y}\, y \;-\; E_{y|x_i,A,B,\Sigma,\mu}\, y)\Gamma^{-1/2} \|^2,$$

where Γ is a positive definite $q \times q$ matrix.

(5.2.14) $IC := 1/n \sum_{i=1}^{n} E_{y|x_i,A,B,\Sigma,\mu} \log\{[y|x_i, Y](y)\}$, where

$[y \mid x, Y]\,(z)$ is the predictive density function evaluated at z.

Naive estimates of these criteria are obtained by using the same data for both the calculation of the predictive distribution and the evaluation of its goodness, i.e. by

(5.2.15) $MSEP_{FIT} := 1/n \sum_{i=1}^{n} \| (E_{y|x_i,Y}\, y \;-\; Y_{i.})\Gamma^{-1/2} \|^2$, and

(5.2.16) $IC_{FIT} := -\, 1/n \sum_{i=1}^{n} \log\{[y|x_i; Y](\, Y_{i.}\,)\}$.

Possibly more realistic estimates of the goodness of the prediction may be obtained by using crossvalidation, i.e. by setting

(5.2.17) $MSEP_{CV} := 1/n \sum_{i=1}^{n} \| (E_{y|x_i,Y_{(i)}}\, y \;-\; Y_{i.})\Gamma^{-1/2} \|^2$, and

(5.2.18) $IC_{CV} := 1/n \sum_{i=1}^{n} \log\{[y|x_i, Y_{(i)}](\, Y_{i.}\,)\}$, where

$Y_{(i)}$ is the $(n-1) \times q$ matrix obtained from Y by elimination of the i–th row $Y_{i.}$.

The crossvalidated predictive distribution $[y|x_i, Y_{(i)}]$ can be used for diagnostic checks of the predictive power, in addition to the predictive distribution $[y|x_i, Y]$.

5.2.7. Rank Determination

In the preceding sections it has always been assumed that the rank r of the coefficient matrix $M = AB'$ in (5.2.1) is known. In many situations it is reasonable to assume a fixed rank, while in other situations the a–priori choice of a fixed rank may be difficult. The choice of the rank r can be regarded as the choice between different models and the standard Bayesian procedure would be to select the model with highest posterior probability (see Gelfand et al., 1992). If the interest is centered on prediction, it is natural to base the choice of the rank r on an adequate prediction criterion, for example those discussed in the preceding section. One would then select the rank

$$r_0 := \arg\min_r \text{MSEP}(r) \quad \text{or} \quad r_0 := \arg\min_r \text{IC}(r) .$$

Alternatively, the selection of the rank can be based on a graphical comparison of the predictive distributions of the different models.

5.3. Markov Chain Monte Carlo Methods

It seems that for reduced rank regression models (5.2.1), an exact analytical derivation of the posterior distribution of the parameters $[A, B, \Sigma, \mu \mid Y]$ is intractable. As an alternative to approximate analytical or difficult numerical integration methods, Markov chain Monte Carlo sampling procedures have been suggested recently (see, for example, Gilks, 1993), based on the "equivalence" between a distribution and samples from this distribution. The outcome of such a Monte Carlo procedure is an approximate i.i.d. sample $\{(A_j, B_j, \Sigma_j, \mu_j); j = 1, ..., N\}$ from a distribution approximating $[A, B, \Sigma, \mu \mid Y]$, where N is a large integer.

5.3.1. Gibbs Sampling of the Posterior Distribution

The Gibbs sampling procedure (see, for example, Gelfand and Smith, 1990) is a natural Monte Carlo procedure for generating a sample from the posterior distribution of $[A,B,\Sigma, \mu \mid Y]$.

Gibbs sampling is defined as follows. Let m (number of sequences) and k (number of iterations) be integers.

Then for $j = 1, \dots, m$, the following steps are performed

Let $A_j^{(0)}, B_j^{(0)}, \Sigma_j^{(0)}, \mu_j^{(0)}$ be arbitrary starting matrices.

Then for $i = 1, \dots, k$,

Sample $\mu_j^{(i)}$ from $[\mu | A_j^{(i-1)}, B_j^{(i-1)}, \Sigma_j^{(i-1)}, Y]$,

Sample $\Sigma_j^{(i)}$ from $[\Sigma | A_j^{(i-1)}, B_j^{(i-1)}, \mu_j^{(i)}, Y]$,

Sample $B_j^{(i)}$ from $[B | A_j^{(i-1)}, \Sigma_j^{(i)}, \mu_j^{(i)}, Y]$,

Sample $A_j^{(i)}$ from $[A | B_j^{(i)}, \Sigma_j^{(i)}, \mu_j^{(i)}, Y]$.

For large enough k, $\{(A_j^{(k)}, B_j^{(k)}, \Sigma_j^{(k)}, \mu_j^{(k)}); j = 1, \dots, m\}$ is then approximately an i.i.d. sample from the posterior distribution $[A, B, \Sigma, \mu \mid Y]$. Hence, if the samples corresponding to the first 1000, say, iterations are ignored and from the rest the samples corresponding to every 10th, say, iteration are taken, an approximately i.i.d. sample $\{(A_j, B_j, \Sigma_j, \mu_j); j = 1, \dots, N\}$ of size $N = m(k-1000)/10$ from $[A, B, \Sigma, \mu \mid Y]$ is obtained.

Note that the distributions from which samples are drawn are constrained and unconstrained matrix normal and inverse Wishart distributions, respectively, given by (5.2.7),(5.2.9),(5.2.10) and either (5.2.8),(5.2.8*) or (5.2.8**). Therefore, drawing of samples from the unconstrained distributions is straightforward (see Odell and Feiveson, 1966). How to sample from the constrained normal distribution is not so obvious. However note that a mapping ϕ from the unconstrained to the constrained space Φ has been introduced in section 5.2.2. An approximate sample from the constrained normal distribution (5.2.10) can simply be obtained by sampling from the normal distribution

$$(5.2.19)\quad N(A^{(1)}, D^{1/2}(B'\Sigma^{-1}B)^{-1}D^{1/2} \otimes {W^{(1)}}^{-1}),$$
$$\text{where } D = (A^{(1)\prime}W^{(1)}A^{(1)})^{1/2}$$

and then mapping the sample by ϕ to the constrained space Φ. When an exact sample from the constrained distribution is needed, the rejection method can be used (see, for example, Smith and Gelfand, 1992), which can be described as follows. For $A \in \Phi$, let $\rho_\Phi(A)$ be the functional form of the constrained normal distribution, i.e. a function which is proportional to the density of the constrained normal distribution. Let $\rho_\phi(A)$ be the functional form of the mapped normal distribution defined above by (5.2.19). If there exists a value M such that $\rho_\Phi(A)/\rho_\phi(A) < M$, then a sample from the constrained normal distribution can be obtained by the following procedure. For any sampled value

A^* from the mapped normal distribution ρ_ϕ, this value is accepted with probability $\rho_\Phi(A^*)/M\rho_\phi(A^*)$. By sampling from the mapped normal distribution until a sampled value is accepted, a sample from the constrained normal distribution is obtained. It remains to calculate the ratio $\rho_\Phi(A)/\rho_\phi(A)$ and the constant M. First note that the functional form $\rho_\Phi(A)$ is just the usual matrix normal density without normalizing constants. Next, by the definition of the mapping ϕ in section 5.2.2,

$$\rho_\phi(A) := \int_{\phi^{-1}(A)} \rho(Z)dZ = \int \rho(FA)dF \ , \ \text{for } A \in \Phi \ ,$$

where ρ is the normal density (5.2.19), F varies over all invertable $r \times r$ matrices and

$\Phi^{-1}(A) := \{ Z \ p \times p \text{ matrix} ; \phi(Z) = A \}$. It is then easy to see that for any matrix normal distribution $N(M, \Omega_0 \otimes W^{-1})$, one has

$$\rho_\phi(A) \propto \exp\{1/2 \ \mathrm{tr}(A'WM\Omega_0^{-1}M'WA)\}$$

and for any normal distribution $N(M, \Omega \otimes W^{-1})$ one has

$$\rho_\Phi(A) \propto \exp\{1/2 \ \mathrm{tr}(A'WM\Omega^{-1} + \Omega^{-1}M'WA)\} .$$

Let $D = (M'WM)^{1/2}$ and $Z = M'D^{-1}WA$. Then one obtains for the ratio

$$\begin{aligned}
&\rho_\Phi(A)/\rho_\phi(A)\\
&\propto \exp\{-1/2 \ \mathrm{tr}(Z'D\Omega_0^{-1}DZ - Z'D\Omega^{-1} - \Omega^{-1}DZ)\}\\
&= \exp\{-1/2 \ \mathrm{tr}((Z - Z_0)'D\Omega_0^{-1}D(Z - Z_0))\}\\
&\quad \cdot \exp\{-1/2 \ \mathrm{tr}(Z_0'D\Omega_0^{-1}DZ_0)\}\\
&\leq M := \exp\{-1/2 \ \mathrm{tr}(Z_0'D\Omega_0^{-1}DZ_0)\} ,
\end{aligned}$$

where $Z_0 = \Omega_0^{-1}D^{-1}\Omega^{-1}$.

Defining $\Omega_0 = D^{1/2}\Omega D^{1/2}$, one then obtains the rejection probability

$$\begin{aligned}
\rho_\Phi(A)/M\rho_\phi(A) &= p(Z)\\
&= \exp\{-1/2 \ \mathrm{tr}((Z - Z_0)'D^{1/2}\Omega^{-1}D^{1/2}(Z - Z_0))\}.
\end{aligned}$$

This result can be used in the present case by setting $M = A^{(1)}$, $W = W^{(1)}$ and $\Omega = (B'\Sigma^{-1}B)^{-1}$.

Note that then $Z = A^{(1)}(A^{(1)\prime}W^{(1)}A^{(1)})^{-1/2}W^{(1)}A$ and $D = (A^{(1)\prime}W^{(1)}A^{(1)})^{1/2}$.

To get a better understanding of the rejection probability, consider a reduced rank regression model of rank 1 with $W^{(1)} = I_p$ and $\Sigma = \sigma^2 I_q$ and define $A_s^{(1)} := A^{(1)}/(A^{(1)\prime}A^{(1)})^{1/2}$. Then it is easy to see that the rejection probability

$$p(A) = \exp\{ -(A^{(1)\prime}A^{(1)})^{1/2}/2\sigma^2 \, (1 - A_s^{(1)\prime}A)^2\} .$$

Hence when the sample A^* is close to the direction of $A^{(1)}$ and/or the variance of the error σ^2 is smaller than the length of $A^{(1)}$, then the sample will be accepted with high probability. In practice these conditions will be fulfilled when the data can be fitted reasonably well by the reduced rank regression model. Hence, the sampling will typically be efficient.

If the distribution of a function of the parameters $f(A, B, \Sigma, \mu)$ is of interest, a sample from the distribution of f is easily obtained as $\{f(A_j, B_j, \Sigma_j, \mu_j); j = 1, ..., N\}$. Statistical summary measures such as the expected value and the expected variance can be calculated by taking the mean and the variance of the sample, respectively. The distribution of the coefficient matrix $M := AB'$ is of special interest for the reduced rank regression models. It should be noted however that while all samples $M_j := A_j B_j'$ are of rank r, its mean $\hat{M} := 1/N \sum A_j B_j'$ has full rank, generally. An alternative more adequate summary measure is the matrix $\tilde{M}$ of rank r defined as

$$\tilde{M} := (1/N \sum A_j)\ (1/N \sum B_j') .$$

5.3.2. Gibbs Sampling of the Predictive Distribution

Let $\{(A_j, B_j, \Sigma_j, \mu_j); j = 1, ..., N\}$ be a sample from the posterior distribution $[A, B, \Sigma, \mu | Y]$. Then for fixed p–dimensional predictor vector x, a sample from the predictive distribution $[y \mid x, Y]$ is easily obtained by
$\{y_j' := \mu_j' + x'A_j B_j + e_j' \ ; j = 1, ..., N\}$, where e_j is sampled from $N(0, \Sigma_j)$.

Alternatively, the predictive distribution itself can be written as a mixture of normal distributions, namely by $1/N \sum N(\mu_j' + xA_j B_j', \Sigma_j)$.

This distribution can be used to estimate the IC criterion for the goodness of prediction.

For the calculation of the MSEP criterion for the goodness of prediction, the expected value of the predictor distribution $[\mu' + x'AB'|x, Y]$ is needed. A sample from this distribution is given by $\{\mu_j' + x'A_jB_j \; ; j = 1, ..., N\}$, so that the mean is an estimate of the required expected value.

If more realistic measures for the goodness of prediction are required, crossvalidation is necessary. Unfortunately, already the estimation of the posterior distribution by Gibbs sampling is time consuming. The crossvalidation procedure, at least in the leave–one–out form, seems therefore only possible for not too large datasets.

5.4. Example

The Bayesian analysis of the QSAR example described in section 1.6 will now be presented in some detail. A prior with small impact on the posterior distribution is used by specifying (see section 5.2.3) :

$$\mu^{(0)\prime} = (0 \;\; 0), \; \lambda^{(0)} = 1, \;\; \Sigma^{(0)} = 10^{-4} I_2, \; c^{(0)} = 1,$$
$$B^{(0)} = 0, \; A^{(0)} = 0, \; W^{(0)} = 10^{-4} X'X .$$

The Gibbs sampling procedure with $m = 5$ sequences and $k = 5000$ iterations gives a sample $\{(A_j, B_j, \Sigma_j, \mu_j); j = 1, ..., N\}$ of size N=2000 (taking only every tenth value and eliminating the first 1000 iterations) from the posterior distribution of $[A, B, \Sigma, \mu \mid Y]$, if the number of iterations k is large enough, i.e. the sampling procedure has converged. To judge the convergence, simple graphical methods are usually proposed (see Gelfand and Smith, 1990, Gelfand et al., 1990, Gilks, 1993), but formal criteria are sometimes used as well (see, for example, Gelman and Rubin, 1992). First the ergodic means of the five sequences are graphically compared, to see if there are drastic differences, possibly due to the multimodality of the posterior distribution. As seen in Figure 5.4.1 this seems not to be the case.

Next, the median and the quartiles for sections of iterations of size 100 are compared, to see how many iterations are needed before the Gibbs sampling procedure has converged. As seen in Figure 5.4.2, the Gibbs sampler seems to converge extremely fast for the considered models. This extremely fast convergence has been noted by Geweke (1993) as well. This is understandable, since in classical statistics the maximum likelihood estimates can be obtained analytically or by an extremely fast converging Gauss–Seidel algorithm (see section 4.3.8).

Figure 5.4.1 *Ergodic means of the five sequences for a parameter (only every tenth value is used). The plots for the other parameters are similar.*

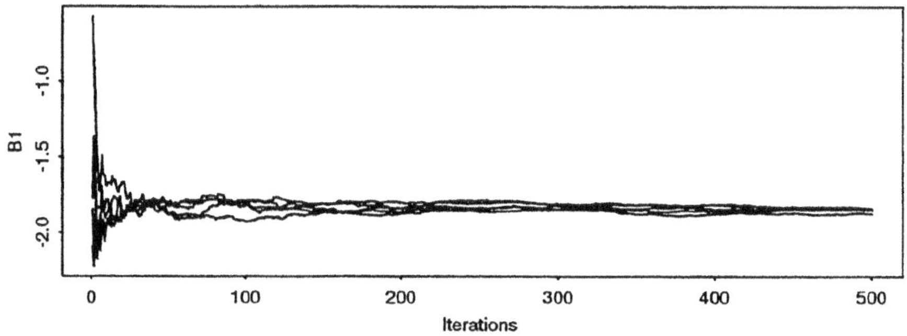

It is easy to approximately calculate the marginal distribution of any parameter, for example by using a kernel density estimator. If quantiles are needed, these can be directly calculated form the sample. In Table 5.4.1 statistical measures summarizing the marginal posterior distribution of the most important parameters are shown. One sees that for the parameter A, mean and median are almost identical and both very close to the maximum likelihood estimate. The standard deviation and the median absolute deviation are very similar as well, so that one location and one scale parameter might be sufficient to describe the distribution of this parameter. The 5% and 95% quantiles enclose 0, except for variables I1 and I2. However, this is misleading, since the predictors are correlated, a point that will be taken up later.

Figure 5.4.2 *Median and quartiles for iteration segments of size 100 for a parameter. The plots for the other parameters are very similar.*

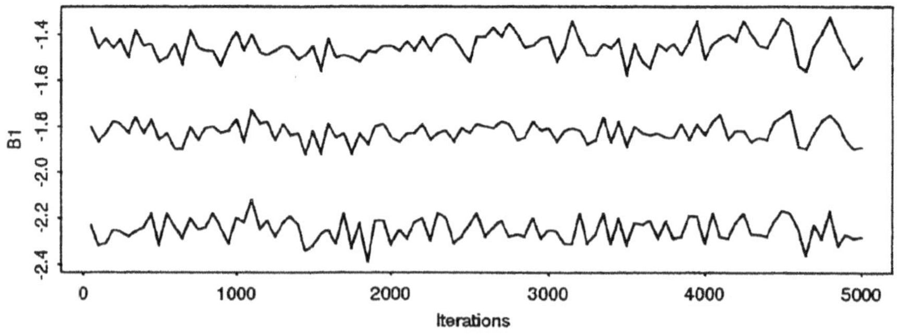

Table 5.4.1 Posterior distribution of the parameter A : summary statistics (mean MEAN, median MEDIAN, standard deviation SDV, median absolute deviation MAD, 5% and 95% quantiles P5, P95)

A	MEAN	MEDIAN	ML	SDV	MAD	P5	P95
I1	0.109	0.112	0.120	0.024	0.025	0.064	0.146
I2	-0.132	-0.135	-0.144	0.023	0.028	-0.170	-0.087
IP23	-0.003	-0.004	-0.006	0.050	0.050	-0.084	0.079
Pi	-0.077	-0.078	-0.081	0.087	0.092	-0.224	0.080
F	0.021	0.022	0.024	0.042	0.043	-0.052	0.089
R	0.062	0.063	0.068	0.058	0.061	-0.036	0.159
MR	-0.065	-0.067	-0.072	0.093	0.095	-0.213	0.091
L	-0.016	-0.015	-0.014	0.109	0.112	-0.199	0.170
B1	0.103	0.106	0.110	0.115	0.116	-0.088	0.285

(ML : Maximum likelihood estimate)

The standard deviations based on classical asymptotic theory (see section 4.3.9) are shown for comparison in Table 5.4.2. However, since these are available for the unusual parametrization I (4.3.4) only, the Bayesian parametrization has to be changed.

Table 5.4.2 Comparison of the standard deviation of the Bayesian posterior distribution and of the classical asymptotic theory for parameter A

	BAYES		MLasy
	SDV	MAD	SDV
I2	1.165	0.342	0.230
IP23	0.946	0.459	0.248
Pi	1.871	0.787	0.575
F	1.059	0.385	0.263
R	1.138	0.582	0.386
MR	1.806	0.901	0.510
L	2.317	0.991	0.651
B1	4.390	1.032	0.759

One sees that the standard deviations based on asymptotic theory are systematically smaller than the corresponding estimates obtained by the Bayesian methodology. The large differences between the standard deviation and the median absolute deviation for this type of parametrization are due to the skewness of the distributions, giving another argument for not using this nonstandard parametrization I (4.3.4) or parametrization II (4.3.5).

For the coefficient matrix M, summary statistics of the posterior distribution are shown in the first part of Table 5.4.3. In the second part of Table 5.4.3 the standard deviations based on classical asymptotic theory are shown. One sees that these standard deviations are smaller than those obtained by a Bayesian analysis.

Table 5.4.3 Posterior distribution of the coefficient matrix M :

ADULTS		MEAN	MEDIAN	SDV	MAD	P5	P95
	I1	-0.226	-0.230	0.064	0.067	-0.331	-0.110
	I2	0.273	0.275	0.064	0.069	0.154	0.382
	IP23	0.006	0.008	0.100	0.100	-0.160	0.165
	Pi	0.158	0.157	0.175	0.189	-0.154	0.475
	F	-0.043	-0.046	0.084	0.087	-0.182	0.102
	R	-0.127	-0.129	0.119	0.122	-0.324	0.072
	MR	0.133	0.137	0.188	0.186	-0.179	0.429
	L	0.033	0.029	0.218	0.224	-0.324	0.402
	B1	-0.212	-0.217	0.231	0.237	-0.587	0.183
LARVAE		**MEAN**	**MEDIAN**	**SDV**	**MAD**	**P5**	**P95**
	I1	-0.214	-0.215	0.063	0.065	-0.316	-0.108
	I2	0.263	0.267	0.075	0.078	0.134	0.385
	IP23	0.014	0.008	0.095	0.096	-0.133	0.178
	Pi	0.143	0.146	0.165	0.179	-0.161	0.420
	F	-0.044	-0.043	0.082	0.084	-0.188	0.091
	R	-0.127	-0.123	0.121	0.122	-0.342	0.061
	MR	0.141	0.130	0.189	0.187	-0.148	0.459
	L	0.019	0.028	0.208	0.215	-0.346	0.362
	B1	-0.201	-0.201	0.222	0.228	-0.566	0.175

Table 5.4.3 (continued) Maximum likelihood estimates and standard deviation based on classical asymptotic theory

M	ML		SDV (MLasy)	
	ADULTS	LARVAE	ADULTS	LARVAE
I1	-0.267	-0.256	0.049	0.051
I2	0.321	0.307	0.052	0.055
IP23	0.013	0.012	0.068	0.065
Pi	0.180	0.173	0.155	0.149
F	-0.054	-0.051	0.072	0.069
R	-0.150	-0.143	0.098	0.094
MR	0.160	0.153	0.135	0.129
L	0.031	0.030	0.177	0.169
B1	-0.244	-0.233	0.206	0.198

The predictive distributions for the test data are summarized in Table 5.4.4. One sees again that the two substances IIia and IIIia behave very differently (compare with section 4.3.10).

Table 5.4.4 Predictions for the test data (outlier substances in bold face)

	ADULTS	mean	5%	95%	LARVAE	mean	5%	95%
IIca	-2.0	-1.723	-2.36	-1.061	>-1.7	-1.789	-2.48	-1.105
IIcb	>-1.7	-2.180	-2.88	-1.474	-2.0	-2.218	-2.96	-1.476
IIia	**-2.6**	**-1.006**	**-2.19**	**0.114**	**-2.6**	**-1.067**	**-2.21**	**0.130**
IIIia	**>-1.7**	**-0.462**	**-1.62**	**0.619**	**-2.0**	**-0.541**	**-1.80**	**0.666**
IIIka	>-1.7	-0.970	-1.62	-0.335	-1.7	-1.076	-1.79	-0.338
IIIla	>-1.7	-1.072	-1.70	-0.478	>-1.7	-1.159	-1.83	-0.494
IIIma	>-1.7	-1.092	-1.78	-0.398	>-1.7	-1.174	-1.95	-0.471

In Figure 5.4.3, the Bayesian predictive distribution for the test data are graphically displayed, as well as the classical asymptotic predictive distribution. One sees that the differences are usually not drastic; however the distributions based on classical asymptotic theory are tighter.

Figure 5.4.3 *Bayesian predictive distribution (–) and classical asymptotic predictive distribution (—) for the test data. The observed response is shown as well (/ or ⊐ for partially incomplete measurements).* ADULTS LARVAE

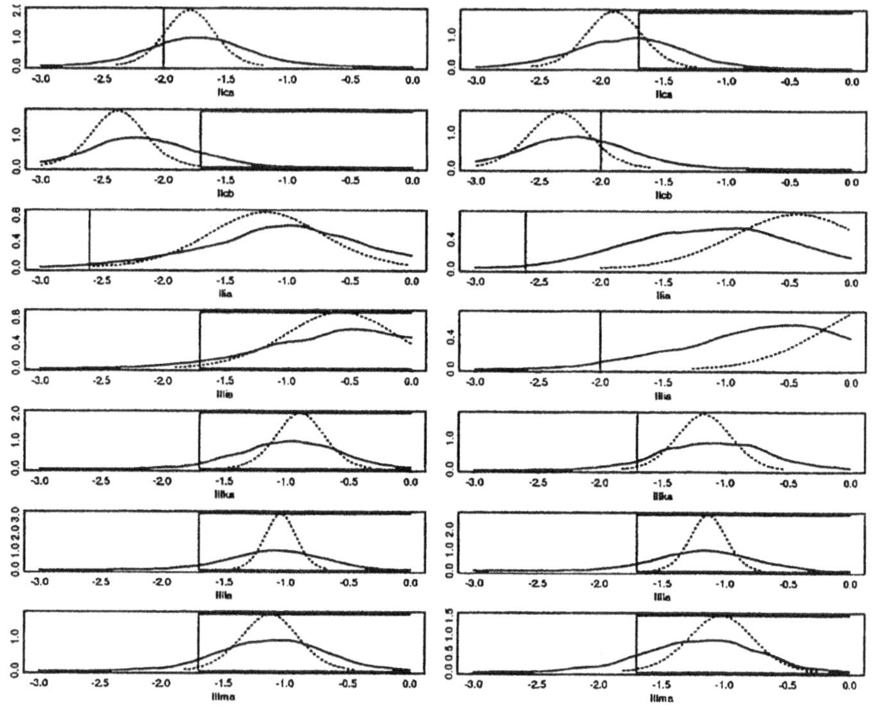

In Figure 5.4.4 the Bayesian predictive distribution for a 1–dimensional reduced rank regression model and for a multivariate regression model are compared for the test data. One sees that the predictive distributions based on the reduced rank regression model are tighter than those obtained by multivariate regression. In addition to these graphical comparisons, the prediction criteria (section 5.2.6) can be calculated as well. As seen on the left side of Table 5.4.5, a full rank model seems to give a somewhat better fit.

Table 5.4.5 Prediction Criteria (FIT, MSEP is mean R^2)

all predictor variables			selected predictor variables		
	DIM1	DIM2		DIM1	DIM2
MSEP	0.676	0.755	MSEP	0.670	0.695
IC	-3.51	-4.06	IC	-3.52	-3.59

Figure 5.4.4 *Bayesian predictive distribution (90% HPD regions) of 1-dimensional reduced rank regression (-) and of multivariate regression (...) for the test data (with observed responses).*

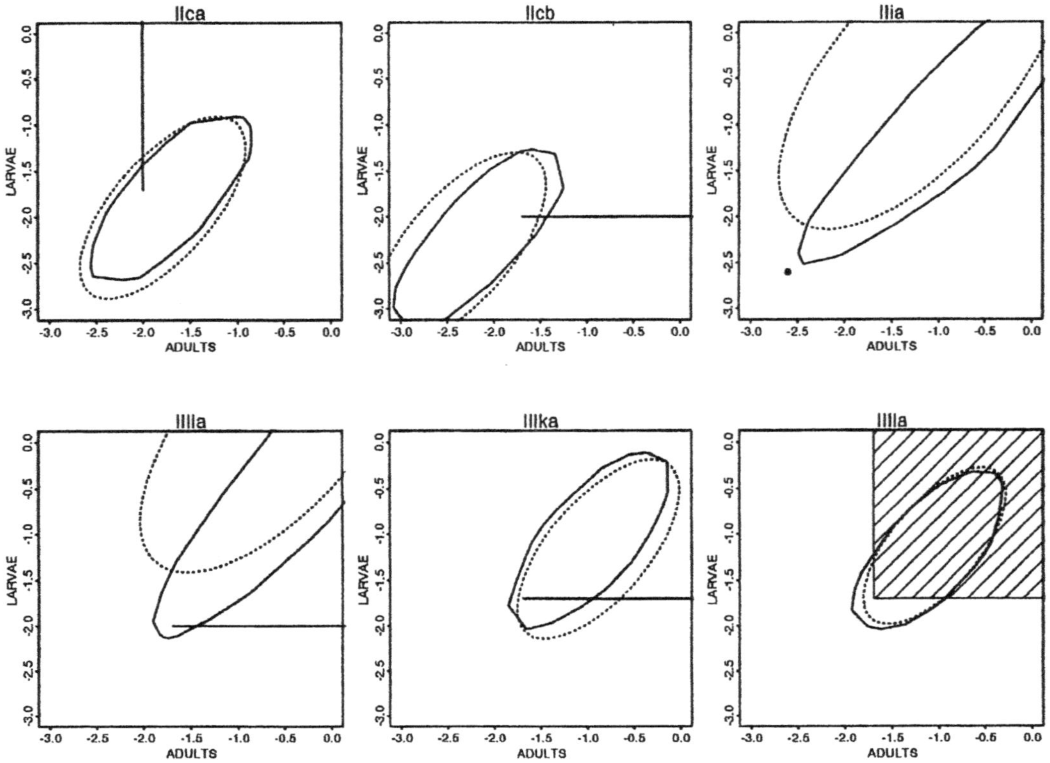

Figure 5.4.5 *Crossvalidation : Bayesian predictive distribution of 1–dimensional reduced rank regression (–) based on reduced data. Additionally, the observed response is shown (|) as well as the maximum–likelihood estimates (⁝) and the crossvalidated maximum likelihood (¦) estimates (for some arbitrarily selected observations).*

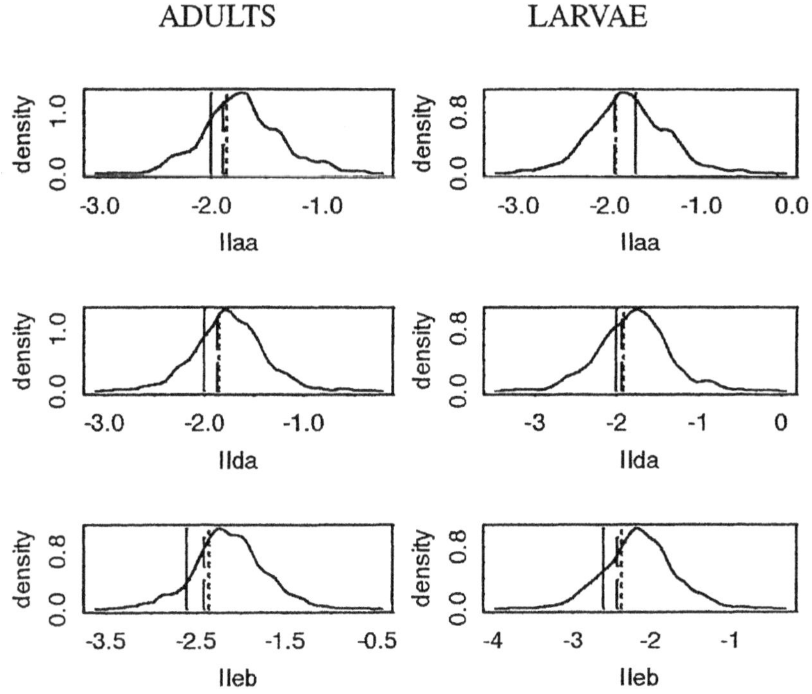

To check the adequacy of the reduced rank regression model, crossvalidation can be used. In Figure 5.4.5, the crossvalidated Bayesian predictive distributions (section 5.3.2) of a 1–dimensional reduced rank regression model are shown, as well as the observed responses, the maximum–likelihood estimates and the crossvalidated maximum likelihood estimates. A comparison with the observed responses suggests that the model is adequate.

As has been mentioned above, the parameter A is difficult to interpret due to the correlation of the predictor variables. Using only the variables selected in the classical forward variable selection procedure (see section 4.6.3), a set of nearly uncorrelated predictor variables is obtained (see Table 5.4.6).

Table 5.4.6 Selected predictor variables
Squared correlations

	I1	I2	F	R	MR
I1	1.000	0.078	0.000	0.001	0.036
I2	0.078	1.000	0.016	0.017	0.006
F	0.000	0.016	1.000	0.011	0.068
R	0.001	0.017	0.011	1.000	0.066
MR	0.036	0.006	0.068	0.066	1.000

The results of a Bayesian Analysis with these selected variables are shown in Table 5.4.7. One sees now, that only the effect of the variable MR might be doubted.

The predictive distributions for the test data are summarized in Table 5.4.8 and Figures 5.4.6 and 5.4.7. One sees that these are very similar to the analysis using all predictor variables. In the right part of Table 5.4.5, the prediction criteria are shown. The full rank model gives only a slightly better fit.

Table 5.4.7 Posterior distribution of the parameters : summary statistics

A	MEAN	MEDIAN	SDV	MAD	P5	P95
I1	0.115	0.117	0.024	0.025	0.072	0.152
I2	-0.126	-0.128	0.019	0.021	-0.156	-0.090
F	0.045	0.045	0.025	0.026	0.002	0.088
R	0.099	0.100	0.021	0.023	0.060	0.135
MR	-0.037	-0.038	0.026	0.027	-0.081	0.007

B	ADULTS	LARVAE
MEAN	-2.112	-1.949
MEDIAN	-2.127	-1.951
SDV	0.291	0.362
MAD	0.311	0.386
P5	-2.575	-2.555
P95	-1.603	-1.323

Table 5.4.8 Predictions for the test data (selected predictor variables only, outlier substances in bold face)

	ADULTS	mean	5%	95%	LARVAE	mean	5%	95%
IIca	-2.0	-1.657	-2.193	-1.103	>-1.7	-1.720	-2.36	-1.085
IIcb	>-1.7	-2.153	-2.698	-1.603	-2.0	-2.175	-2.84	-1.531
IIia	**-2.6**	**-0.686**	**-1.349**	**-0.042**	**-2.6**	**-0.823**	**-1.58**	**-0.099**
IIIia	**>-1.7**	**-0.151**	**-0.827**	**0.511**	**-2.0**	**-0.325**	**-1.14**	**0.444**
IIIka	>-1.7	-1.036	-1.548	-0.535	-1.7	-1.150	-1.77	-0.509
IIIla	>-1.7	-1.070	-1.627	-0.511	>-1.7	-1.175	-1.80	-0.525
IIIma	>-1.7	-1.268	-1.846	-0.683	>-1.7	-1.364	-2.02	-0.689

Figure 5.4.6 *Selected variables : Bayesian predictive distribution (–) and classical asymptotic predictive distribution (—) for the test data. The observed response is shown as well (/ or ⊐ for partially incomplete measurements).*

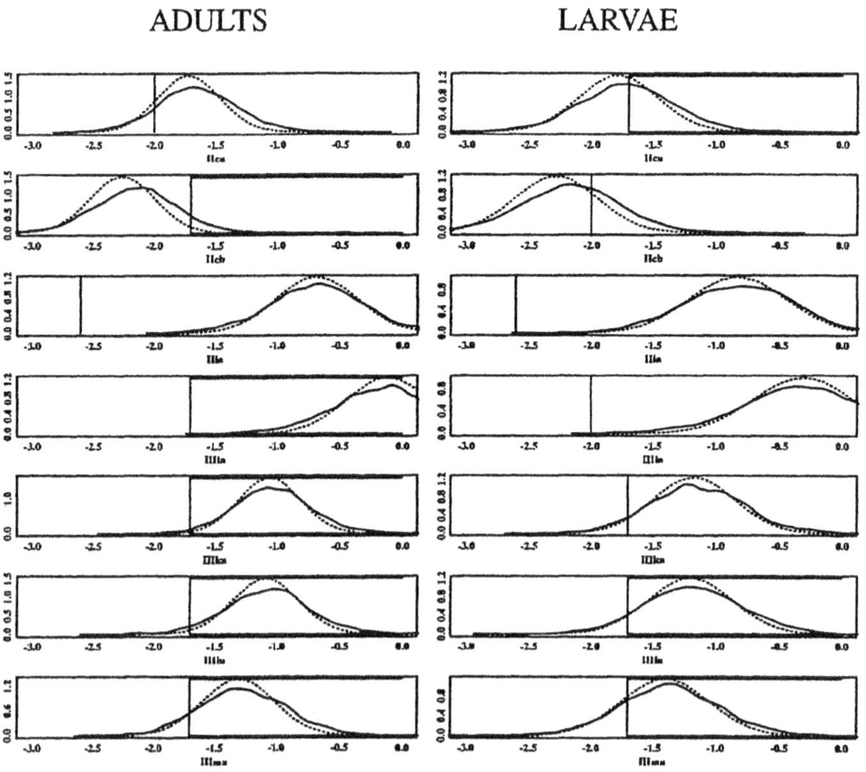

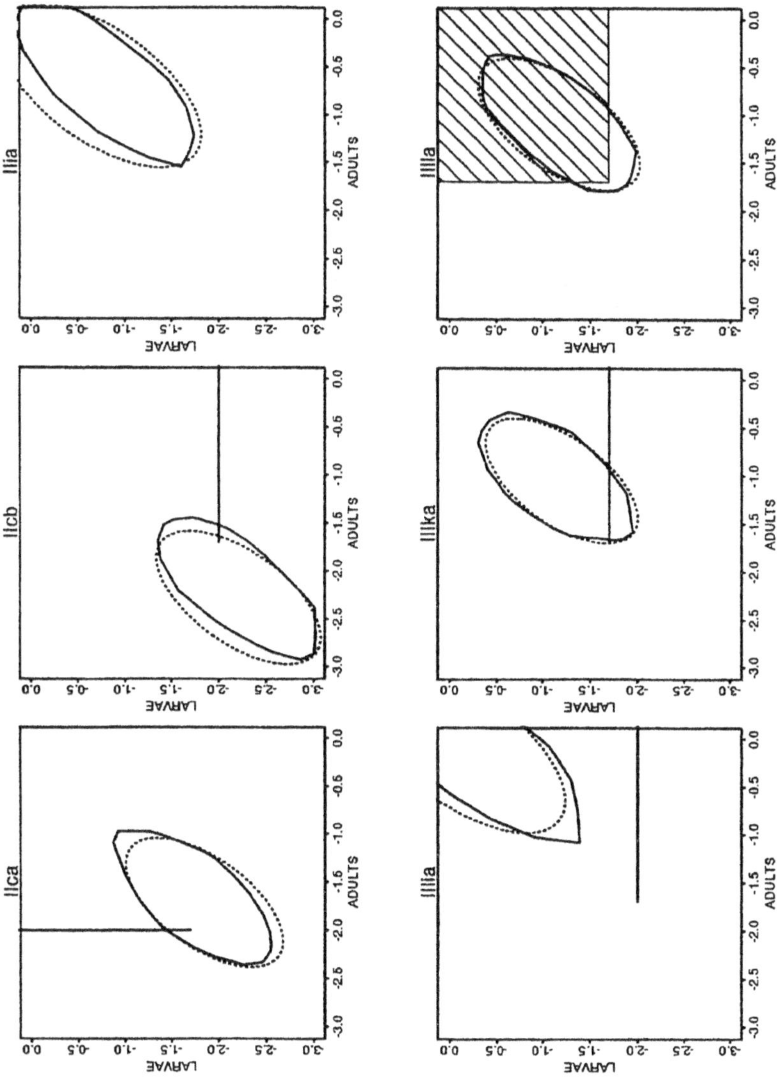

Figure 5.4.7 *Selected variables : Bayesian predictive distribution (90% HPD regions) of 1-dimensional reduced rank regression (-) and of multivariate regression (...). The observed response is shown as well.*

6.

Case Studies

6.1. ®Voltaren : An Anti–Inflammatory Drug

6.1.1. Data

®Voltaren (diclofenac) is a very efficient anti–inflammatory drug developed by CIBA–GEIGY Ltd.. In a retrospective study, analogs with even higher potency are searched for. 32 substances were synthesized as derivatives of diclofenac. The modifications were done by substituting the H–atoms at 6 different positions of the basic molecular structure in the following way (see Figure 6.1.1) :

- At each of the 5 positions of benzene A (P2–P6), the H–atom is either replaced by one of the substituents Cl, Br, F, OH, Me, OMe, Et, or left unchanged;
- At one position of benzene B (P5'), the H–atom is either replaced by a one of the substituents Cl, F, OH, or left unchanged;

To describe the structure of these substances, the following physico–chemical parameters are used.

- logP : measure of the lipophilicity of the substance (experimental values);
- Sigma : measure of the electronic properties of the substance; sum of Hammett sigma constants of the substituents of benzene A (see Hansch and Leo, 1979),

Figure 6.1.1 *The basic structure of diclofenac, consisting of two benzenes, A and B. The H–atoms in position P2–P6 and P5' are substituted to obtain derivatives.*

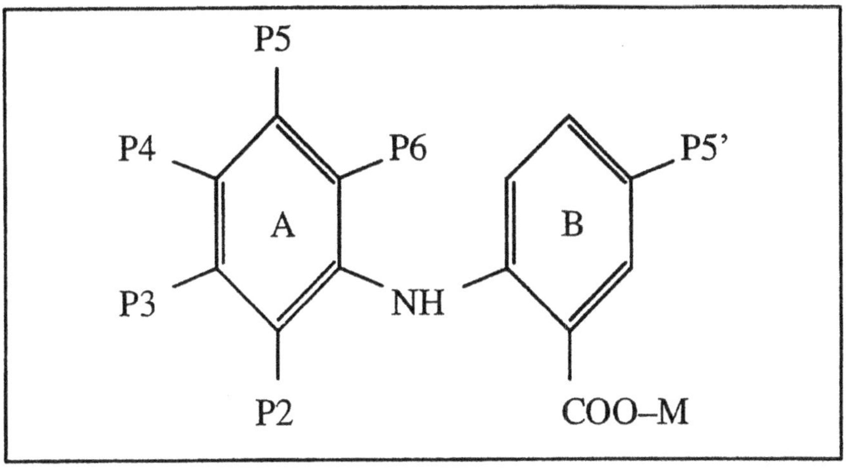

- Alpha : angle between benzene A and B of the substance (experimental values, partially interpolated);
- B1, B5 : descriptors of the shape of the ortho–substituents of benzene A (P2,P6), namely the sum of the smallest and largest width (see Verloop et al., 1976).

Since the substances are putative anti–inflammatory drugs, their biological activity is defined as the ability of the substance to inhibit inflammation. An in–vivo and an in–vitro test are designed to measure this activity. The in–vitro test measures the concentration IC50 needed to inhibit 50% of a specific enzyme relevant for inflammation (cyclooxygenase). The corresponding response variable is called PGS and defined as log(1/IC50). The in–vivo test measures the effective dose ED40 needed to diminish the paw volume of rats with adjuvant arthritis by 40%. The corresponding response variable is called AdA and is defined as log(1/ED40).

The two response variables are obtained from completely independent experiments, so that the errors are certainly independent. However, the size of the errors cannot be assumed to be equal for the two response variables. The biologists are confident that the mechanism causing the anti–inflammatory effect is the inhibition of the specific enzyme used in the in–vitro test. Hence, it is reasonable to assume that the response

Table 6.1.1 Voltaren data

Learning Dataset

	P2	P3	P4	P5	P6	P5′	PGS	AdA	logP	SIGMA	ALPHA	B1	B5
S1	Cl	H	F	H	Cl	F	0.25	0.22	1.62	2.62	69	1.80	3.6
S2	Cl	H	H	H	Cl	F	0.13	0.02	1.48	2.56	69	1.80	3.6
S3	Cl	H	H	H	Cl	Cl	0.10	0.04	1.97	2.56	69	1.80	3.6
S4	Cl	H	H	H	Cl	H	0.00	-0.01	1.13	2.56	69	1.80	3.6
S5	Cl	H	OMe	H	Cl	H	0.00	-0.49	1.24	2.29	69	1.80	3.6
S6	Cl	H	H	H	Me	H	0.00	-0.74	1.17	1.57	69	1.80	3.8
S7	Cl	H	F	H	Cl	H	-0.05	-0.47	1.27	2.62	69	1.80	3.6
S8	Cl	H	H	H	Br	H	-0.28	-0.17	1.44	2.63	72	1.95	3.7
S9	Me	Me	H	H	H	H	-0.30	-1.59	1.03	0.22	62	1.52	3.0
S10	Cl	H	H	H	I	H	-0.30	-0.11	1.64	2.62	75	2.15	3.9
S11	Me	Me	H	Me	Me	H	-0.78	-1.25	1.95	0.44	69	1.52	4.0
S12	Me	H	H	H	Me	H	-0.78	-0.29	1.21	0.58	69	1.52	4.0
S13	Cl	OH	H	H	Cl	H	-1.00	-1.51	0.72	2.68	69	1.80	3.6
S14	Me	Cl	H	H	H	H	-1.04	-0.86	1.54	0.66	62	1.52	3.0
S15	Cl	OMe	H	H	Cl	H	-1.04	-0.96	1.18	2.68	69	1.80	3.6
S16	Cl	Me	H	H	Cl	H	-1.15	-0.68	1.69	2.49	69	1.80	3.6
S17	Cl	Me	H	H	H	H	-1.30	-0.26	1.59	1.21	62	1.80	2.8
S18	Cl	H	H	H	F	H	-1.75	-0.80	0.56	2.21	62	1.80	3.1

S19	Me	Me	H	Me	Me	Cl	-1.82	-1.20	2.93	0.44	69	1.52	4.0
S20	Et	H	H	H	Et	H	-1.83	-1.55	1.91	0.82	69	1.52	6.3
S21	Cl	H	Cl	H	H	H	-1.85	-1.23	1.33	1.51	62	1.80	2.8
S22	Cl	OMe	OH	H	Cl	H	-1.86	-2.47	0.51	2.31	69	1.80	3.6
S23	H	Cl	H	Cl	H	H	-1.90	-2.23	1.98	0.74	58	1.00	2.0
S24	Cl	H	H	Cl	H	H	-2.00	-1.01	1.64	1.65	62	1.80	2.8
S25	F	H	F	H	H	H	-2.01	-2.88	0.55	0.99	60	1.35	2.3
S26	Cl	H	OH	H	Cl	H	-2.15	-0.98	0.70	2.19	69	1.80	3.6
S27	H	Cl	Cl	H	H	H	-2.15	-2.53	1.80	0.60	58	1.00	2.0
S28	Cl	H	H	H	H	H	-2.31	-1.76	0.91	1.28	62	1.80	2.8
S29	F	H	H	H	F	H	-2.36	-2.28	0.26	1.86	60	1.35	2.7
S30	H	H	H	H	H	H	-2.46	-2.94	0.19	0.00	58	1.00	2.0
S31	Cl	OH	OMe	H	Cl	H	-2.47	-2.47	0.25	2.41	69	1.80	3.6
S32	Cl	H	OH	H	Cl	OH	-3.04	-2.78	0.27	2.19	69	1.80	3.6

Test Dataset

	P2	P3	P4	P5	P6	P5′	PGS	AdA	logP	SIGMA	ALPHA	B1	B5
S33	Cl	H	H	H	Cl	OH	-3.35	-0.98	0.76	2.52	69	1.90	3.6
S34	H	CF3	H	H	H	H	-2.48	-2.53	1.47	0.43	58	1.00	2.0
S35	Cl	Cl	H	H	H	H	-1.00	NA	1.66	1.65	62	1.80	2.8

PGS,AdA : biological activity (NA : no data available)
logP,SIGMA,ALPHA,B1,B5 : physico-chemical properties

variables are closely related. A scatterplot of the two variables suggests that the two responses are approximately linearly related (see Figure 6.1.2), so that a one–dimensional reduced rank regression model (4.3.7) with diagonal error covariance matrix (RRR, see section 4.3.8.4) seems to be adequate.

As a reference, the whole learning dataset is shown in Table 6.1.1 as well as some additionally synthesized test substances with partially incomplete biological information. A detailed description of this example is given in Moser et al. (1990).

Figure 6.1.2 *In–vitro test (PGS) and in–vivo test (AdA) of the anti–inflammatory effect of diclofenac analogs. A diagonal line is added.*

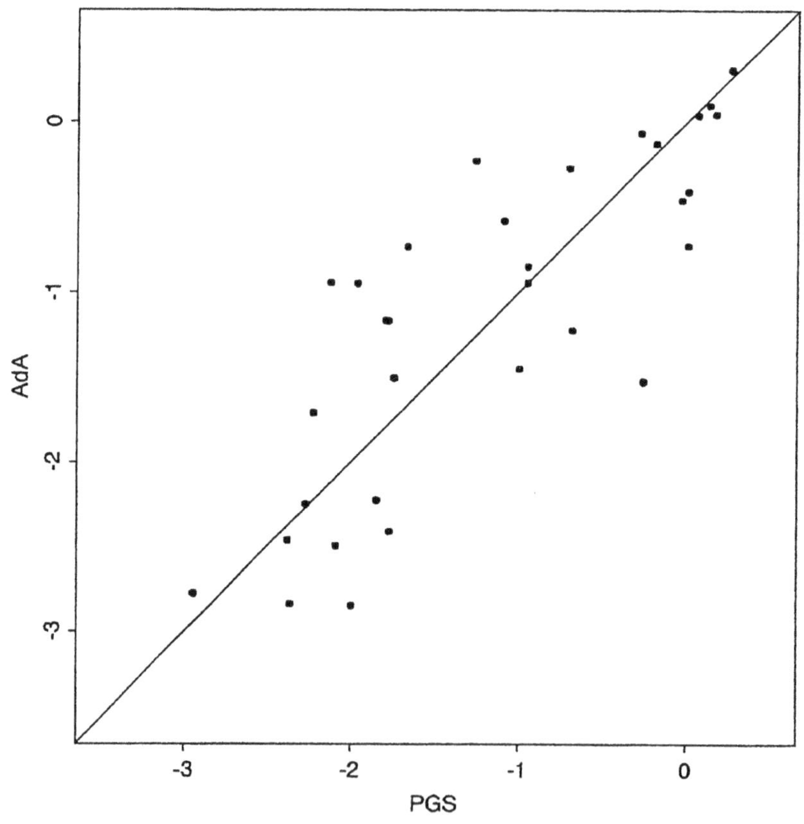

Figure 6.1.3 *Crossvalidation predictions of the 1–dimensional reduced rank regression model with diagonal error covariance matrix (RRR) : 2 outliers.*

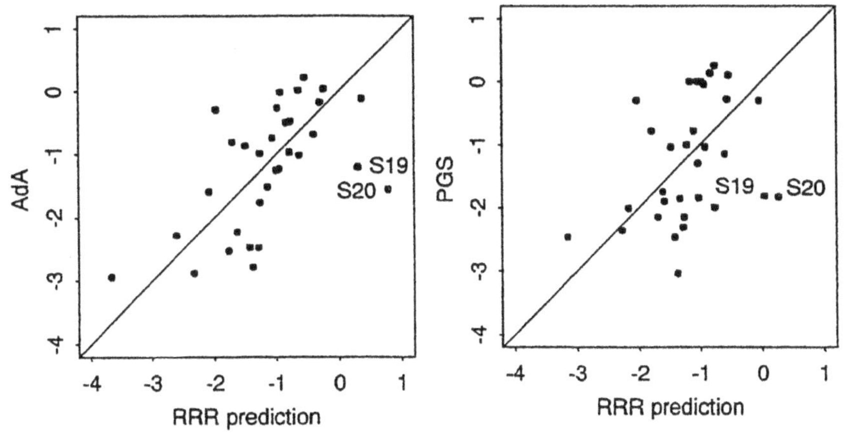

6.1.2. Analysis

A reduced rank regression model (4.3.7) of rank 1 with diagonal error covariance matrix (RRR, see section 4.3.8.4) seems to be adequate to describe the relationship between predictors and responses. The cross-validation predictions of this model shown in Figure 6.1.3 indicate that the substances S20 and S19 are behaving different than the other substances. Hence these substances are added to the test set.

As a reference, the predictive power of the heuristic multivariate prediction methods (see chapter 3) principal component regression PCR and partial least squares PLS is estimated.

Table 6.1.2 Estimation of the prediction criteria MSEP is mean R^2 (maximal values in bold)

PCR	DIM1	DIM2	DIM3	DIM4	DIM5
MSEP	0.329	0.600	0.610	0.624	0.644
MSEPasy	0.300	**0.543**	0.524	0.510	0.502
MSEPcv	0.251	**0.524**	0.455	0.430	0.399

PLS	DIM1	DIM2	DIM3	DIM4	DIM5
MSEP	0.482	0.603	0.624	0.631	0.644
MSEPcv	0.413	**0.488**	0.449	0.401	0.399

Table 6.1.2 shows that a 2–dimensional PCR prediction model and a 2–dimensional PLS prediction model seem to be optimal.

Next, the predictive power of the reduced rank regression model without (RRR) and with (RRRv) variable selection is estimated. These estimates in Table 6.1.3 show that a 1–dimensional reduced rank regression model gives better or only slightly worse predictions than multivariate regression. The estimates of the predictive power for variable selection procedures suggest that 2 or at most 3 variables are needed in the model. The steric parameter B5 and the measure of lipophilicity, logP, seem to be the most important variables.

The coefficient matrices of the four models are shown in Table 6.1.4, as well as the prediction model for RRRv. While PLS and PCR are very similar, there are considerable differences to RRR.

Table 6.1.3 Estimation of the prediction criteria
MSEP is mean R^2 (2.3.2)
(maximal values in bold face)

RRR	DIM1	DIM2
MSEP	0.618	0.644
MSEPasy	**0.535**	0.502
MSEPcv	0.398	**0.399**
IC	-0.112	-0.242
ICasy	**0.421**	0.558
ICCV	0.849	**0.841**

RRRv	B5	logP	B1	ALPHA	SIGMA
MSEP	0.327	0.592	0.608	0.610	0.618
MSEPasy	0.298	0.549	**0.552**	0.540	0.534
MSEPcv	-0.140	**0.454**	0.387	0.375	0.397
IC	1.089	0.078	-0.041	-0.064	-0.112
ICasy	1.356	0.411	**0.358**	0.402	0.421
ICCV	2.142	**0.666**	0.886	0.924	0.850

Table 6.1.4 Coefficient matrices of the prediction models

M

	PCR		PLS		RRR		RRRv	
	PGS	AdA	PGS	AdA	PGS	AdA	PGS	AdA
CONST	-1.195	-1.182	-1.195	-1.182	-1.195	-1.182	-1.195	-1.182
logP	0.505	0.516	0.507	0.523	0.491	0.581	0.493	0.528
SIGMA	0.034	0.060	0.036	0.061	0.145	0.171	0.000	0.000
ALPHA	0.183	0.215	0.151	0.183	-0.329	-0.390	0.000	0.000
B1	0.117	0.147	0.129	0.159	0.166	0.196	0.000	0.000
B5	0.174	0.204	0.193	0.224	0.567	0.670	0.502	0.539

Prediction model for RRRv

$$PGS = -1.20 + 0.70\ Z$$
$$AdA = -1.18 + 0.76\ Z$$

$$Z = 0.70\ logP + 0.71\ B5$$

The predictions of the test data are shown in Table 6.1.5. One notes that the outliers S19 and S20 are in fact extremely badly predicted. A very bad prediction is observed for the PGS value of substance S33, while its AdA value is very well predicted. This indicates, that this substance shows drastically different behavior in the in–vivo and in–vitro test.

Table 6.1.5 Predictions for the test data (badly predicted values in bold face)

	OBSERVED		PCR		PLS		RRR		RRRv	
	PGS	AdA	PGS	AdA	PGS	AdA	PGS	AdA	PGS	AdA
S33	**-3.35**	-0.98	**-1.21**	-1.12	**-1.21**	-1.12	**-1.17**	-1.15	**-1.26**	-1.25
S34	-2.48	-2.53	-1.92	-2.13	-1.93	-2.14	-2.15	-2.31	-1.96	-2.01
S35	-1.00	NA	-0.97	-0.99	-0.95	-0.96	-0.83	-0.76	-1.13	-1.11
S19	**-1.82**	**-1.20**	**0.64**	**0.69**	**0.64**	**0.71**	**0.56**	**0.89**	**1.01**	**1.18**
S20	**-1.83**	**-1.55**	**0.39**	**0.55**	**0.46**	**0.63**	**1.88**	**2.46**	**2.02**	**2.26**

Figure 6.1.4 *Crossvalidation predictions for the four models PCR, PLS, RRR and RRRv : substances with extreme deviations are marked.*

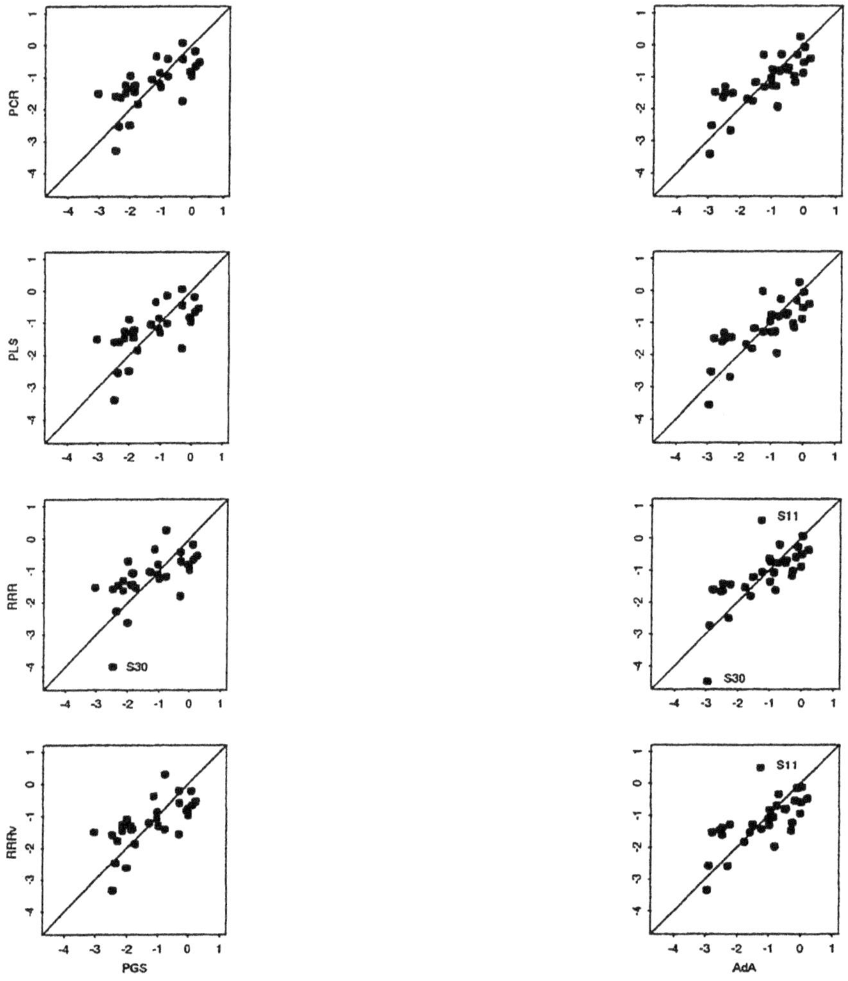

The four models are graphically compared in Figure 6.1.4. One sees that RRR but not RRRv has difficulties predicting the biological activity of substance S30, and both have difficulties predicting the PGS value of substance S11, which has a very similar chemical structure as the initially identified outlier S19.

Figure 6.1.5 *Ergodic means of the five sequences for an arbitrarily selected parameter.*

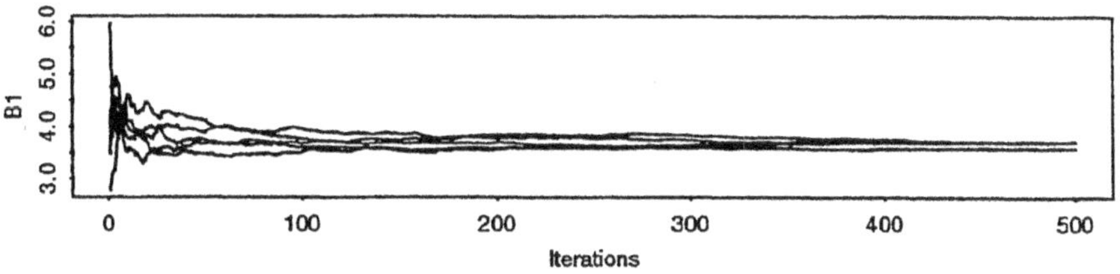

Figure 6.1.6 *Median and quartiles for iteration segments of size 100 for a parameter.*

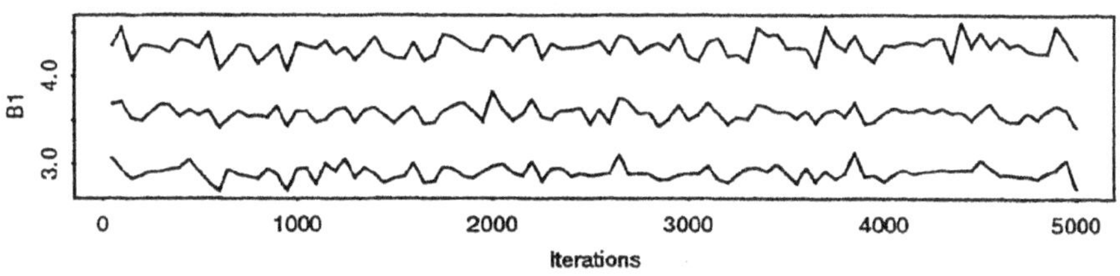

A Bayesian analysis of the data using the Gibbs sampling procedure is performed according to chapter 5. As can be seen from the ergodic means and the medians and quantiles in Figures 6.1.5 and 6.1.6, the convergence is again very fast.

The results of the analysis are summarized in Table 6.1.7. The posterior distribution of the parameter A suggests that the variables logP and B5 are important for prediction, a result already obtained by classical variable selection. Note, however, that the variable ALPHA, and to a lesser extend the variables SIGMA and B1 are correlated with B5, hence interpretations requires great care.

Table 6.1.6 Squared correlations of the predictor variables

	logP	SIGMA	ALPHA	B1	B5
logP	1.0000	0.0029	0.0131	0.0018	0.0072
SIGMA	0.0029	1.0000	0.4836	0.5671	0.3515
ALPHA	0.0131	0.4836	1.0000	0.5844	0.8903
B1	0.0018	0.5672	0.5845	1.0000	0.5711
B5	0.0072	0.3515	0.8904	0.5711	1.0000

Table 6.1.7 Posterior distribution of the parameters : summary statistics

A	MEAN	MEDIAN	SDV	MAD	P5	P95
logP	0.125	0.126	0.015	0.015	0.099	0.149
SIGMA	0.037	0.037	0.033	0.034	-0.021	0.092
ALPHA	-0.083	-0.084	0.068	0.071	-0.199	0.034
B1	0.042	0.044	0.037	0.038	-0.022	0.101
B5	0.145	0.147	0.064	0.066	0.036	0.248

B	PGS	AdA
MEAN	3.721	4.396
MEDIAN	3.741	4.387
SDV	0.749	0.596
MAD	0.762	0.599
P5	2.456	3.428
P95	5.016	5.356

M

PGS	MEAN	MEDIAN	SDV	MAD	P5	P95
logP	0.466	0.464	0.107	0.113	0.292	0.663
SIGMA	0.136	0.135	0.120	0.130	-0.073	0.351
ALPHA	-0.305	-0.297	0.245	0.273	-0.769	0.119
B1	0.153	0.154	0.138	0.143	-0.084	0.382
B5	0.538	0.527	0.264	0.270	0.126	0.992
AdA	**MEAN**	**MEDIAN**	**SDV**	**MAD**	**P5**	**P95**
logP	0.551	0.551	0.097	0.100	0.387	0.717
SIGMA	0.162	0.160	0.143	0.152	-0.087	0.418
ALPHA	-0.367	-0.365	0.304	0.317	-0.900	0.135
B1	0.187	0.187	0.169	0.169	-0.094	0.457
B5	0.636	0.632	0.294	0.299	0.154	1.138

The predictive distributions for the test data are summarized in the first part of Table 6.1.8 and Figure 6.1.7.

Figure 6.1.7 *Bayesian predictive distribution of 1-dimensional reduced rank regression (-) and of multivariate regression (...) for the test data. The observed response is shown as well.*

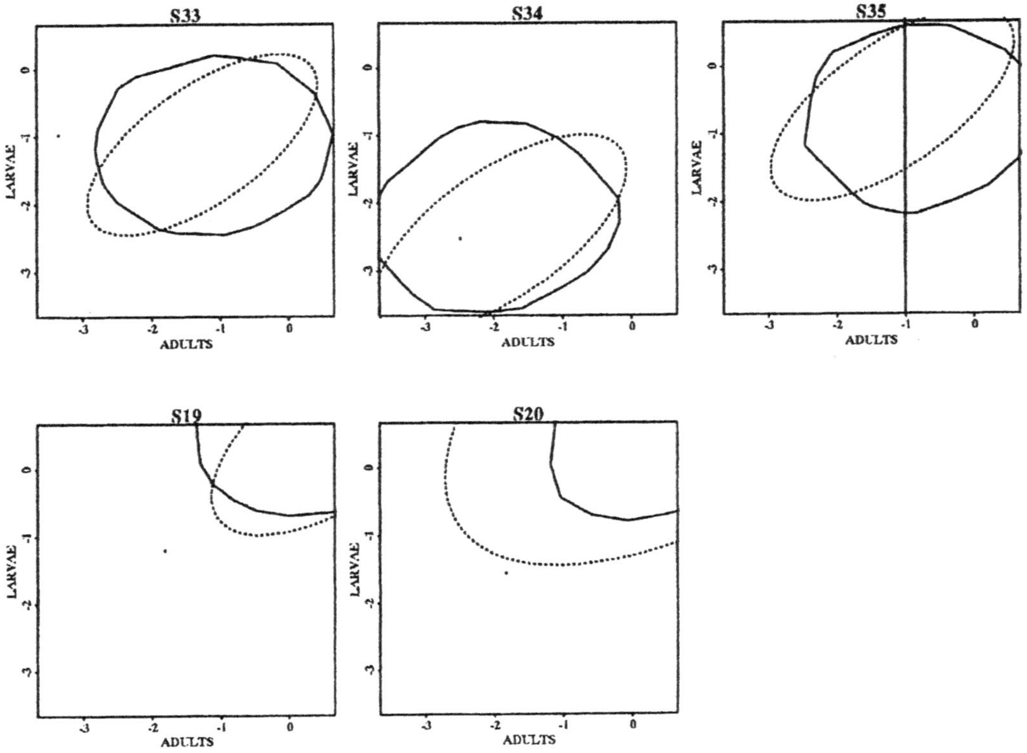

Again, the PGS value of substance S33 and the outliers S19 and S20 cannot be well predicted. The prediction criteria in the second part of Table 6.1.8 suggest that the full rank model gives only a slightly better fit than the reduced rank regression model.

Table 6.1.8 Predictions for the test data
(badly predicted values in bold face)

	PGS	mean	5%	95%	AdA	mean	5%	95%
S33	**-3.35**	**-1.128**	**-2.423**	**0.151**	**-0.98**	**-1.114**	**-2.158**	**-0.123**
S34	-2.48	-2.044	-3.355	-0.673	-2.53	-2.223	-3.302	-1.179
S35	-1.00	-0.819	-2.099	0.486	NA	-0.745	-1.802	0.300
S19	**-1.82**	**0.538**	**-0.917**	**1.972**	**-1.20**	**0.833**	**-0.384**	**2.040**
S20	**-1.83**	**1.804**	**-0.491**	**4.236**	**-1.55**	**2.317**	**-0.111**	**4.835**

Prediction criteria (FIT)

	DIM1	DIM2
MSEP	0.615	0.644
IC	-0.09	-0.24

6.2. Development of a Herbicide

6.2.1. Data

A putative lead substance has been the starting point for the synthesis of 30 derivatives (chemical substances) in CIBA–GEIGY Ltd. with the aim to get a new herbicide. The modifications of the putative lead structure were obtained by substituting one position of the lead structure by different other substituents.

Figure 6.2.1 *Growth inhibiting effect of the putative herbicides on five different weeds : scatterplot matrix of biological activity. Diagonal lines are added.*

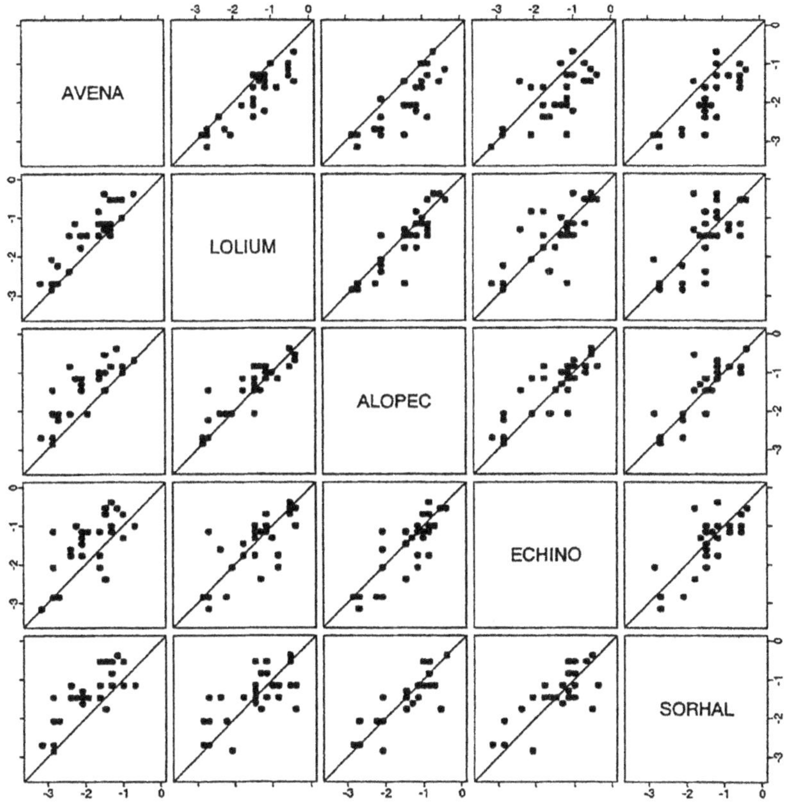

Figure 6.2.2 *Growth inhibiting effect of the putative herbicides on five different weeds : boxplot of biological activity.*

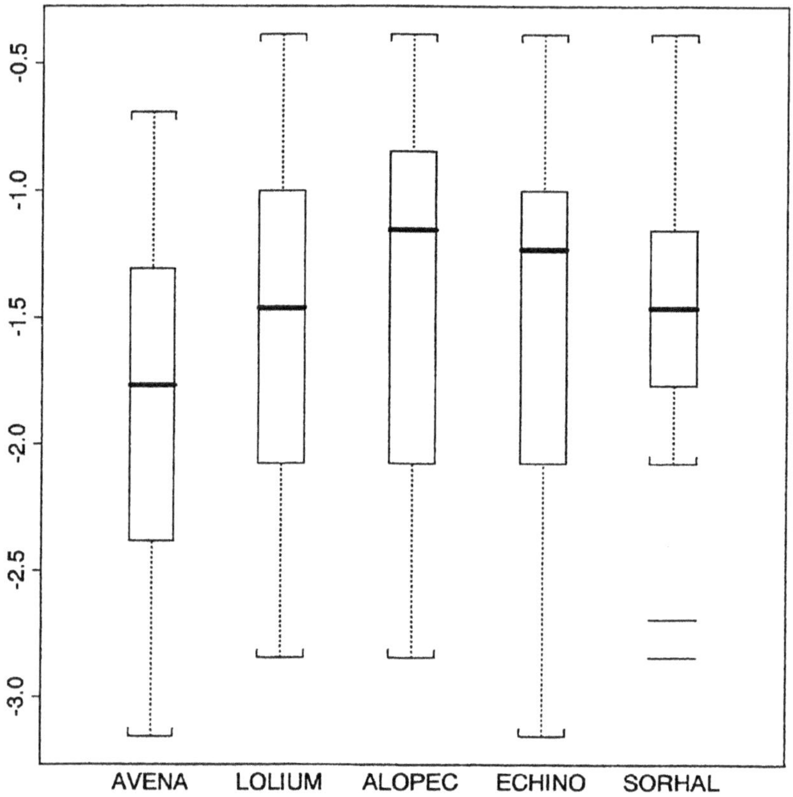

The substances are intended to be used as herbicides. Hence their biological activity is defined as their ability to inhibit weed growth. The reaction of five weeds of genus *Avena*, *Lolium*, *Alopecurus*, *Echinochloa* and *Sorghum* on the substance is measured for each weed separately as the concentration IC50 inhibiting 50% of the growth of the weed. The response variables are called AVENA, LOLIUM, ALOPEC, ECHINO and SORHAL, respectively, and are defined as log(1/IC50) for each of the corresponding weeds. The logarithms are taken to get approximate homoscedasticity of the measurement errors (see section 2.2).

The five response variables are obtained in a common experiment, so that some dependence of the errors seems possible. The biologists are confident that the mechanism causing the growth inhibiting effect is the same in all weeds, hence it is reasonable to assume that the response variables are closely related (compare the discussion in section 4.2). A scatterplot matrix of the variables suggests that even a linear relationships might exist between the five responses (see Figure 6.2.1). However some weeds are obviously much more sensitive than others as can be seen better in Figure 6.2.2, where parallel boxplots of the responses are shown.

To describe the structure of these substances, the following physico–chemical descriptors are used.

- logP : measure of the lipophilicity of the chemical substance, thought to be important for the transport of the herbicide to the weed (see, for example, Hansch and Leo, 1979);
- MR : measure of the overall size of the chemical substance (see, for example, Hansch and Leo, 1979);
- DM, IP, LUMO : measures of the electronic properties of the chemical substance, namely the Dipole Moment (DM), the Ionization Potential (IP) and the Lowest Unoccupied Molecular Orbit energy (LUMO), calculated by approximate semi–empirical quantum–mechanical methods (MOPAC software; see, for example, Dewar and Thiel, 1977).

From Figures 6.2.1 and 6.2.2 and the description of the experiment it seems appropriate to assume a reduced rank regression model (4.3.7) of rank 1 with unstructured error covariance matrix (CCA, canonical correlation analysis, see section 4.3.8.3).

As a reference, information about the whole learning dataset can be found in Table 6.2.1.

6.2.2. Analysis

Since no reasonable fit with a linear model of the predictor variables was obtained, a full quadratic model is used in the following. First the heuristic multivariate prediction methods (see chapter 3) principal component regression PCR and partial least squares PLS are used. The estimates of their predictive power in the first part of Table 6.2.2 suggest the use of a 4–dimensional PCR and a 2–dimensional PLS method.

Table 6.2.1 Herbicide data

	AVENA	LOLIUM	ALOPEC	ECHINO	SORHAL	DM	IP	LUMO	logP	MR
S1	−2.846	−2.846	−2.692	−2.846	−2.077	6.32	9.62	−0.967	3.871	10.57
S2	−2.846	−2.077	−2.077	−2.077	−2.846	5.56	9.93	−0.848	3.192	10.42
S3	−1.615	−0.846	−1.154	−2.077	−1.462	5.58	9.83	−1.053	0.082	9.71
S4	−1.308	−0.538	−0.846	−0.385	−1.154	4.71	10.74	−1.434	2.249	8.42
S5	−2.077	−1.769	−1.462	−1.462	−1.462	4.42	10.44	−1.401	0.273	10.15
S6	−1.462	−1.308	−1.462	−2.385	−1.769	3.88	10.45	−1.416	−0.086	10.33
S7	−1.923	−1.462	−2.077	−1.154	−1.462	3.87	9.10	−0.740	1.093	8.28
S8	−1.154	−0.538	−0.385	−0.538	−0.385	4.93	10.37	−1.019	2.284	9.48
S9	−1.615	−1.462	−1.154	−1.154	−1.462	4.74	10.53	−1.496	0.090	10.19
S10	−1.308	−1.154	−0.846	−1.000	−0.846	4.68	9.59	−1.271	2.457	10.12
S11	−1.308	−1.462	−0.846	−1.000	−0.538	4.99	9.47	−1.160	2.126	9.18
S12	−1.462	−1.154	−1.000	−0.692	−0.538	4.36	10.45	−1.394	0.288	9.71
S13	−2.077	−1.462	−1.308	−1.308	−1.615	6.34	9.69	−0.849	2.061	9.21
S14	−2.846	−2.846	−2.846	−2.846	−2.692	3.66	10.43	−1.343	0.223	10.79
S15	−2.077	−1.769	−1.154	−1.769	−1.462	4.27	10.44	−1.391	3.184	10.24
S16	−2.385	−2.385	−2.077	−1.615	−1.462	6.08	9.94	−0.809	3.164	10.20
S17	−3.154	−2.692	−2.692	−3.154	−2.692	6.19	9.19	−1.001	1.810	10.08
S18	−2.077	−1.462	−1.462	−1.154	−1.308	6.44	9.58	−0.826	2.739	10.13
S19	−2.231	−1.154	−1.154	−1.000	−1.462	4.20	8.92	−0.710	2.348	9.21

S20	-1.462	-0.385	-0.538	-0.538	-1.769	4.26	10.48	-1.066	2.138	8.56
S21	-1.615	-1.154	-1.000	-1.154	-0.538	4.25	10.26	-0.888	1.442	9.61
S22	-2.385	-1.462	-0.846	-1.769	-1.154	3.83	9.55	-0.950	0.762	9.62
S23	-1.308	-1.308	-0.846	-1.154	-0.846	5.33	10.01	-0.972	2.247	8.96
S24	-1.000	-0.538	-0.846	-0.692	-0.538	4.94	10.40	-1.039	1.944	9.01
S25	-2.692	-2.692	-2.231	-2.846	-2.077	6.45	9.53	-0.804	3.414	10.42
S26	-1.615	-0.846	-1.154	-1.769	-1.154	6.00	9.76	-0.858	0.269	9.71
S27	-2.692	-2.231	-2.077	-2.846	-2.077	4.81	9.07	-0.821	1.544	9.86
S28	-1.000	-1.000	-1.000	-1.308	-1.154	6.86	10.58	-1.326	1.125	9.93
S29	-0.692	-0.385	-0.692	-1.000	-1.154	6.30	10.21	-0.956	1.254	9.43
S30	-2.846	-2.692	-1.462	-1.154	-1.462	4.00	9.56	-0.963	1.291	10.08

AVENA,LOLIUM,ALOPEC,ECHINO,SORHAL : biological activity

DM, IP, LUMO, logP, MR : physico-chemical properties

The reduced rank regression model (4.3.7) with unstructured error covariance matrix (canonical correlation analysis, CCA, see section 4.3.8.3) is very instable, as seen from the crossvalidation estimates in the second part of Table 6.2.2. If the variable selection procedure is used, a much more stable model is obtained. It seems that the overall size MR and the ionization potential IP are by far the most important variables.

Table 6.2.2 Estimated prediction criteria
MSEP is mean R^2 (maximal values in bold)

PCR

	DIM1	DIM2	DIM3	DIM4	DIM5	DIM6	DIM7	DIM8
MSEP	0.062	0.083	0.114	0.519	0.532	0.558	0.570	0.583
MSEPasy	0.046	0.050	0.066	0.454	0.451	**0.461**	0.456	0.453
MSEPcv	-0.111	-0.133	-0.255	**0.296**	0.213	0.206	0.190	0.140

PLS

	DIM1	DIM2	DIM3	DIM4	DIM5	DIM6	DIM7	DIM8
MSEP	0.339	0.501	0.522	0.530	0.540	0.575	0.630	0.665
MSEPcv	0.161	0.275	**0.281**	0.274	0.163	0.204	0.017	0.112

CCA

	DIM1	DIM2	DIM3	DIM4	DIM5
MSEP	0.643	0.800	0.854	0.890	0.918
MSEPasy	0.498	0.622	0.624	0.611	0.594
MSEPcv	-0.940	-0.403	-0.703	-0.862	-0.913
IC	-8.023	-10.987	-12.613	-13.678	-14.464
ICasy	-5.423	-6.920	-7.213	-7.078	-6.798
ICCV	-2.661	-2.846	-0.988	-0.094	0.076

CCAv

	MR*MR	IP*MR	MR	DM*DM	LUMO*logP	IP*LUMO
MSEP	0.349	0.520	0.665	0.656	0.627	0.628
MSEPasy	0.332	0.496	**0.635**	0.618	0.582	0.577
MSEPcv	0.237	0.346	0.502	**0.523**	0.481	0.397
IC	-5.237	-5.664	-6.051	-6.091	-6.225	-6.225
ICasy	-3.903	-4.264	-4.584	-4.558	**-4.625**	-4.559
ICCV	-4.496	-4.658	-4.858	-5.022	**-5.053**	-4.862

In Table 6.2.3, the coefficient matrix is shown for the reduced rank regression model in combination with variable selection (CCAv), as well as the prediction equation for CCAv. The reduced rank regression model is much easier to interpret than, for example PCR or PLS, were the coeffcient matrix contains over 100 parameters.

Ignoring the unstable method CCA, the three methods PCR, PLS and CCAv are graphically compared in Figure 6.2.3. CCAv is clearly superior to both PCR and PLS.

Table 6.2.3 Coefficient matrix of CCA with variable selection

M CCAv					
	AVENA	LOLIUM	ALOPEC	ECHINO	SORHAL
CONST	-1.903	-1.503	-1.380	-1.528	-1.421
MR	7.378	8.895	7.955	8.285	6.614
MR*MR	-8.127	-9.798	-8.762	-9.125	-7.285
IP*MR	0.451	0.544	0.486	0.506	0.404

Maximum likelihood estimates of the parameters of the reduced rank regression model (CCAv)

	AVENA	LOLIUM	ALOPEC	ECHINO	SORHAL
μ	-1.90	-1.50	-1.38	-1.53	-1.42

	AVENA	LOLIUM	ALOPEC	ECHINO	SORHAL
B	7.38	8.90	7.95	8.28	6.61

A	DIM1
MR	1.01
MR*MR	-1.10
IP*MR	0.06

Figure 6.2.3 *Crossvalidation predictions for the three methods PCR, PLS and CCAv.*

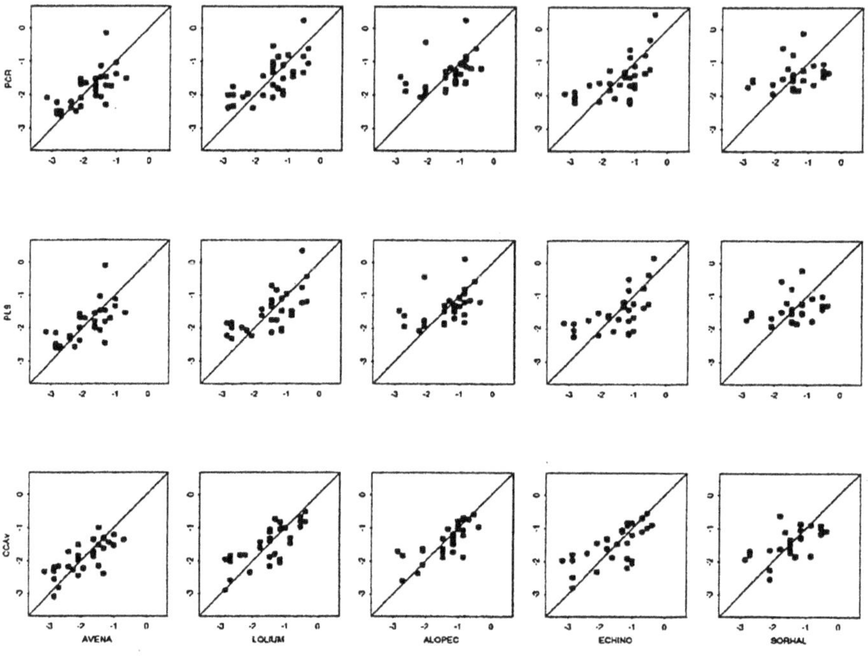

The data are analyzed as well by Bayesian methods using the Gibbs sampler (see chapter 5). A full quadratic model with the variables MR and IP is used. As seen from Table 6.2.4, the predictors are not very highly correlated, which will make the interpretation of the coefficients easier.

Table 6.2.4 Squared correlations of the predictor variables

	IP	IP*IP	MR	MR*MR	MR*IP
IP	1.000	0.000	0.001	0.011	0.072
IP*IP	0.000	1.000	0.115	0.053	0.015
MR	0.001	0.115	1.000	0.000	0.008
MR*MR	0.011	0.053	0.000	1.000	0.011
MR*IP	0.072	0.015	0.008	0.011	1.000

Figure 6.2.4 *Ergodic means of the five sequences for an arbitrarily selected parameter. The plots for all the other parameters are very similar.*

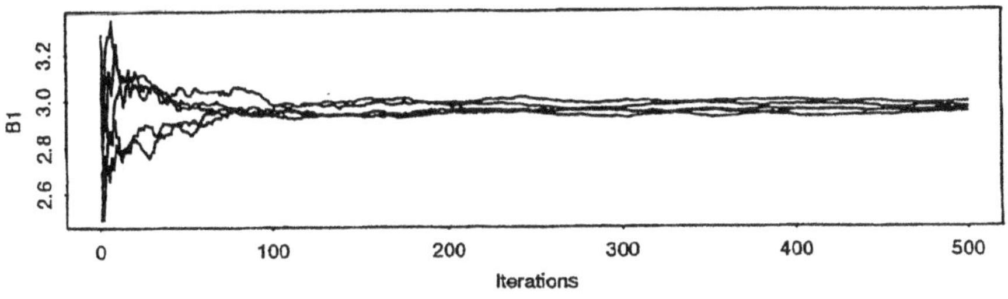

Figure 6.2.5 *Median and quartiles for iteration segments of size 100 for a parameter.*

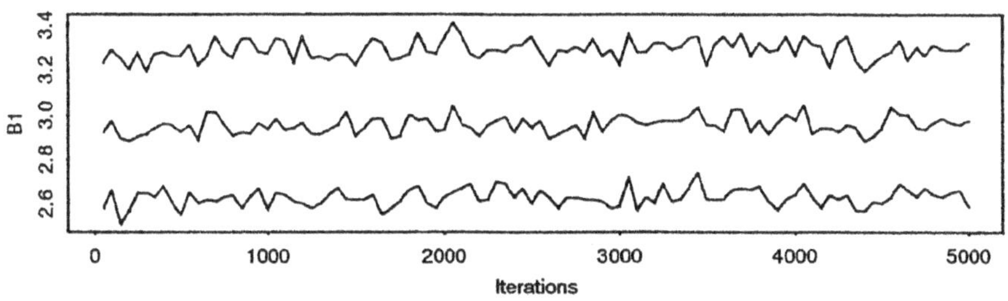

Again, this Markov chain Monte Carlo method is converging extremely fast, as can be clearly seen from the plots of the ergodic means in Figure 6.2.4 and the medians and quantiles in Figure 6.2.5.

The main results of the Bayesian analysis are summarized in Table 6.2.5. Only the effect of the quadratic term in IP seems doubtful, since the 90% HPD interval of A for the corresponding element just encloses zero, while this is not the case for all other marginal intervals. The scaling differences do not seem to be important, since all 90% HPD intervalls for B are strongly overlapping. The prediction criteria in the second part of Table 6.2.5 indicate that a reduced rank regression model of rank 1 is not worse than the full model

Table 6.2.5 Posterior distribution of selected parameters : summary statistics

A

	MEAN	MEDIAN	SDV	MAD	P5	P95
IP	0.575	0.578	0.044	0.050	0.489	0.650
IP*IP	-0.073	-0.073	0.048	0.053	-0.158	0.015
MR	-0.746	-0.750	0.039	0.043	-0.810	-0.670
MR*MR	-0.494	-0.491	0.052	0.061	-0.592	-0.401
MR*IP	0.481	0.487	0.224	0.241	0.096	0.861

B

	AVENA	LOLIUM	ALOPEC	ECHINO	SORHAL
MEAN	2.955	3.490	3.098	3.297	2.587
MEDIAN	2.959	3.477	3.099	3.308	2.592
SDV	0.414	0.470	0.352	0.536	0.487
MAD	0.429	0.489	0.372	0.566	0.512
P5	2.232	2.715	2.499	2.351	1.725
P95	3.654	4.292	3.705	4.205	3.433

Prediction Criteria

	DIM1	DIM5
MSEP	0.667	0.541
IC	-6.02	-5.07

The predictive distributions for a selected part of the learning data are graphically shown in Figure 6.2.7. The comparison with the observed values suggests that the model is adequate.

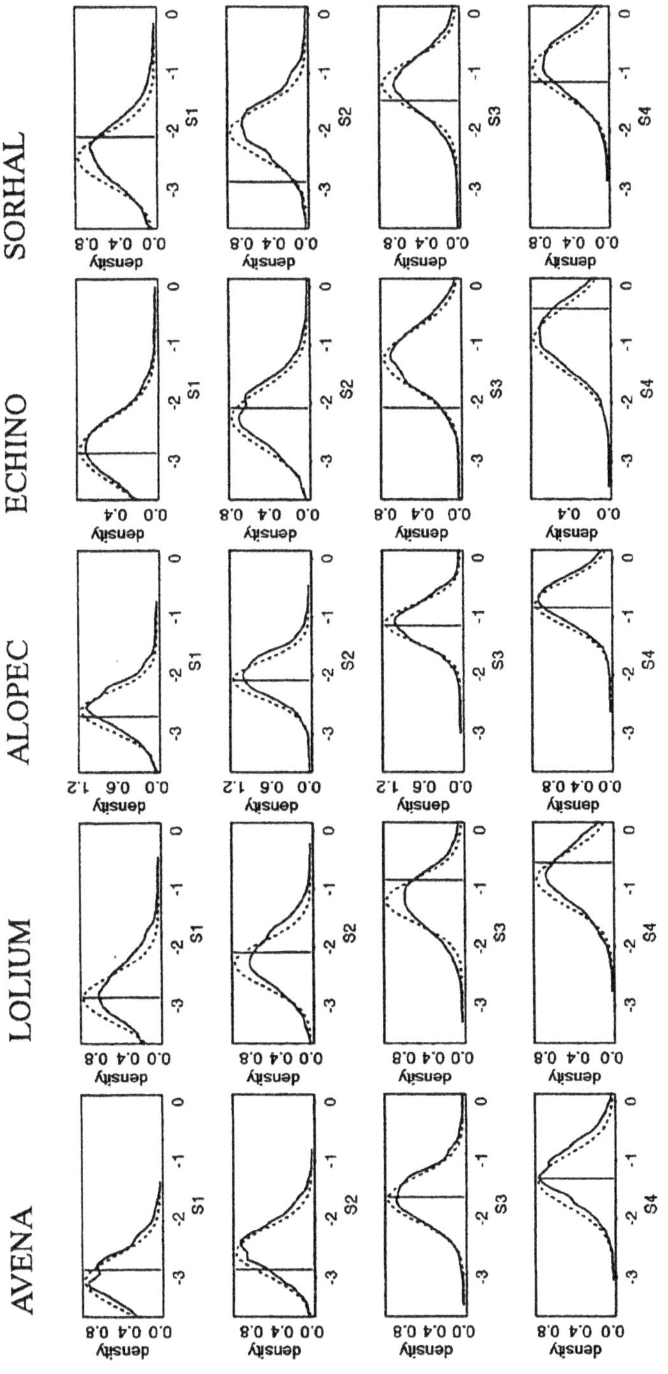

Figure 6.2.6 *Bayesian predictive distribution (–) and classical asymptotic predictive distribution (...) for a selected part of the learning data. The observed response is shown as well (|).*

7.
Discussion

The prediction of multivariate responses by multivariate predictors is an important problem in applied statistics. In real applications, a "true model" for the link between responses and predictors is never available. However, some basic insight into the application area helps to suggest statistical models which are often useful to obtain some understanding about the relationship between responses and predictors. In the area of QSAR (quantitative structure–activity relationships), where structural properties of chemical substances are related to their effect on biological systems, reduced rank regression models are very natural (see section 4.2). That these can provide in fact a good approximation for real data can be seen in the case studies presented above (see, for example, sections 4.5.4, 4.6.3 and chapter 6). Heuristic methods such as principal component regression and partial least squares are often used in QSAR studies. They usually provide as good approximations as the reduced rank regression models, but the interpretation of the outcomes of these methods is typically more difficult; see, for example, section 6.2.2. The interaction between chemical substances and biological systems is very complex and the proposed linear multivariate methods such as the reduced rank regression models cannot possibly approximate this relationship in a completely satisfactory way. Hence there is still a lot of scope for the development of more refined nonlinear methods for QSAR and other application areas (compare the discussion in section 4.4.4).

The usefulness of a statistical method in real applications is of primary concern to those who want to understand a specific problem. However, while the behavior of such models in a number of real case studies should be known, it is necessary and instructive to investigate their behavior in idealized situations, that is when the data is assumed to be in fact generated by the considered statistical model. For example, asymptotic theory helps to understand in which situations reduced rank regression models may be expected to give more precise predictions than multivariate regression for large sample sizes (see section 4.3.9), and simulation studies give insight into various aspects of the models for small sample sizes (see sections 4.5.2.2, 4.5.3.2 and 4.6.2). It would be interesting to know as well the reaction of the models to various types of misspecifications to obtain a better understanding of what problems might occur when the methods are applied to real data. Further research is necessary to study some of these aspects.

In QSAR studies, usually only a small number of substances are synthesized for which physico–chemical properties and biological activities are available. It is therefore important that the data is not overfitted, i.e. that a parsimonious model is selected. Since the prediction of biological activity is the main aim, methods have been looked for which are simple without loosing any relevant predictive power. For reduced rank regression models, simplicity has been obtained by a reduction of the dimensionality and of the number of predictor variables. The loss of predictive power has been assessed both by numerical criteria and by graphical methods, hopefully guaranteeing that no oversimplification occurs.

The question of how to adequately assess a statistical prediction method is still an area of intensive research in both theoretical and applied statistics. The numerical criteria and graphical methods discussed in chapter 2 and used in chapters 3 and 4 are certainly not the conclusive answer to this important question. However, both real case studies (see chapter 6) and simulation studies (see sections 4.5 and 4.6) seem to suggest that these procedures are useful for the search of simple prediction models.

There are two fundamentally different approaches to statistical inference, the classical or frequentist approach and the Bayesian approach. While both are consistent mathematical theories, there is still an ongoing controversy among statisticians, which approach is valid in real applications. A large part of this treatise has been concerned with the classical analysis of reduced rank regression in chapter 4 : maximum likelihood estimates for different error covariance structures and the asymptotic distribution of the maximum likelihood estimates have been derived, and the predictive power of the models has been assessed by prediction criteria and graphical methods. The main problem with such a classical analysis is the difficulty to obtain exact confidence regions for the parameters and prediction (tolerance) regions for the responses, so that one has often to rely on asymptotic results (see section 4.3.9).

In chapter 5, a Bayesian analysis of reduced rank regression has been presented. A Bayesian analysis of such a complex multivariate model would until recently have required either approximations similar to those needed in classical analysis or very difficult high–dimensional numerical integration. However the recent recognition that Markov chain Monte Carlo methods, especially the Gibbs sampler (Gelfand and Smith, 1990), can be used for almost exact Bayesian inference has drastically changed the situation. Through the specification of the full conditional posterior distributions of the reduced rank regression model, and the use of an adequate rejection method for random sampling from these distributions, the joint posterior distribution could be derived. In contrast to a classical analysis, where only asymptotic approximations are available, corresponding Bayesian highest posterior density regions for the parameters can be obtained which are almost exact. Once the posterior distribution is available, the calculation of the predictive distribution is straightforward in a sampling context. Based on this predictive distribution, all aspects of Bayesian prediction could be considered : the prediction of a multivariate response for given multivariate predictor, the check of the adequacy of a prediction model, and the selection of the most appropriate prediction model among a set of candidates.

Appendix

A.1 Introduction

Multivariate linear prediction methods are calculated from an $n \times p$ matrix X of predictors and an $n \times q$ matrix Y of responses. The methods in sections A.3 – A.7 all calculate a $p \times d^*$ matrix L_X of loadings. The q–dimensional predicted response $\hat{y}$ for given p–dimensional predictor vector x is then obtained by a multivariate regression of the orthonormal scores $Z = (X - 1_n m_X')L_X$ on the responses Y, using the method described in section A.2. More precisely

(A.1.1) $\hat{y}' = m_Y' + (x - m_X)'M$, where
$M = L_X L_X'(X - 1_n m_X')'(Y - 1_n m_Y')$ ist the $p \times q$ coefficient matrix,
$m_Y = 1/n \ Y'1_n$ is the mean of the matrix Y and
$m_X = 1/n \ X'1_n$ is the mean of the matrix X.

In the following, ε is a small value, $\varepsilon = 5e{-}8$, say.

A.2 Multivariate Regression MR

MR

INPUT : X, Y;

OUTPUT : M.

Let X be an $n \times p$ matrix, Y an $n \times q$ matrix of real values. A multivariate regression (MR) of X and Y gives a $p \times q$ coefficient matrix M.

The coefficient matrix is computed as follows. The truncated singular value decomposition of the centered matrix X gives

(A.2.1) $(X - 1_n m_X') \approx U_X S_X V_X'$, with
$n \times r_X$ column orthogonal matrix U_X, $p \times r_X$ column orthogonal matrix V_X and diagonal $r_X \times r_X$ matrix S_X with $S_{X11} \geq S_{X22} \geq \ldots \geq S_{Xr_Xr_X}$.

r_X is the maximal integer, such that the diagonal elements of S_X are all greater than $\varepsilon\, S_{X11}$; if S_{X11} is 0, then MMR cannot be applied.

The coefficient matrix M is then calculated as

(A.2.2) $M = V_X S_X^{-1} U_X'(Y - 1_n m_Y')$.

If X is known to be orthonormal, then the coefficient matrix can be more easily calculated as

(A.2.3) $M = (X - 1_n m_X')'(Y - 1_n m_Y')$.

A.3 Principal Component Analysis PCA

PCA

INPUT : X, d, v_X;

OUTPUT : L_X, c.

Let X be an $n \times p$ matrix and let the dimension d be a positive integer with $d \leq \min(p, (n - 1))$. Let v_X be an indicator variable, which is equal to 1 if the data are to be scaled (PCA–COR) and equal to 0 otherwise (PCA–COV).

A principal component analysis (PCA) of X gives loadings and eigenvalues. The loadings are a $p \times d^*$ matrix L_X and the eigenvalues a d^*–dimensional vector c , where $d^* \leq d$.

Loadings and eigenvalues are computed as follows. Let m_X be the mean of the matrix X .

If $v_X = 1$, let the diagonal $p \times p$ matrix

$D_X = \mathrm{diag}(\ 1/n\ (X - 1_n m_X')'(X - 1_n m_X')\ )$

be the variance of the matrix X; if $v_X = 0$, define D_X as the $p \times p$ unit matrix I_p.

The truncated singular value decomposition of the centered and scaled matrix X then gives

(A.3.1) $(X - 1_n m_X')D_X^{-1/2} \approx U_X S_X V_X'$, with
$n \times d^*$ matrix U_X, $p \times d^*$ matrix V_X and diagonal $d^* \times d^*$ matrix S_X, where $d^* = \min(d, r)$.

r is the maximal integer, such that the diagonal elements of S_X are all greater than εS_{X11}; if S_{X11} is 0, then PCA cannot be applied.

The loadings and eigenvalues are then defined as

(A.3.2) $L_X = D_X^{-1/2} V_X S_X^{-1}$,
$c = \mathrm{diag}(S_X)$.

A.4 Partial Least Squares PLS

PLS

INPUT : X, Y, d, v_X, v_Y;

OUTPUT : L_X, L_Y, c.

Let X be an $n \times p$ matrix, Y an $n \times q$ matrix of real values. Let the dimension d be a positive integer with $d \le \min(p, (n-1))$. Let v_X be an indicator variable, which is equal to 1 if the X matrix is to be scaled and equal to 0 otherwise. Let v_Y be an indicator variable, which is equal to 1 if the Y matrix is to be scaled and equal to 0 otherwise. The methods are called PLS–COVX–COVY (v_X=0, v_Y=0), PLS–CORX–COVY (v_X=1, v_Y=0), PLS–COVX–CORY (v_X=0, v_Y=1) and PLS–CORX–CORY (v_X=1, v_Y=1).

A partial least squares analysis (PLS) between X and Y gives loadings and eigenvalues. The loadings are a $p \times d^*$ matrix L_X and a $q \times d^*$ matrix L_Y, with $d^* \le d$ and the eigenvalues are a d^*–dimensional vector c. Loadings and eigenvalues are computed as follows. Let m_X and m_Y be the means of the matrices X and Y.

If $v_X = 1$, define the diagonal $p \times p$ matrix D_X as the variance of the matrix X; if $v_X = 0$, define D_X as the $p \times p$ identity matrix I_p. If $v_Y = 1$, define the diagonal $q \times q$ matrix D_Y as the variance of the matrix Y; if $v_Y = 0$, define D_Y as the $q \times q$ unit matrix I_q.

The centered and scaled matrices are then defined as

$$\text{(A.4.1)}\quad \tilde{X} = (X - 1_n m_X{}')D_X^{-1/2} \quad \text{and} \quad \tilde{Y} = (Y - 1_n m_Y{}')D_Y^{-1/2}\ .$$

Then, for $i = 1, 2, \ldots, d^* \leq d$, the following iterative procedure is applied.

The first dimension of the singular value decomposition of $\tilde{X}'\tilde{Y}$ gives

(A.4.2a) $U_i S_i V_i{}'$, with
p–dimensional vector U_i, q–dimensional vector V_i and scalar S_i .

If S_1 is 0 , then $d^* := 0$ and PLS cannot be applied.

If S_i is $0, i > 1$, then $d^* := i - 1$ and the iteration ends; otherwise the matrices $\tilde{X}$ and $\tilde{Y}$ are redefined as

$$\text{(A.4.2b)}\quad \tilde{X} = \tilde{X} - \tilde{U}_i\tilde{U}_i{}'\tilde{X} \quad \text{and} \quad \tilde{Y} = \tilde{Y} - \tilde{U}_i\tilde{U}_i{}'\tilde{Y} \quad \text{, where}$$

$$\tilde{U}_i = (\tilde{X}U_i) / \sqrt{(\tilde{X}U_i)'(\tilde{X}U_i)}\ .$$

After the end of the iterative procedure, define the orthogonal loadings $\tilde{L}_X$ and non–orthogonal scores $\tilde{S}_X$ as

$$\text{(A.4.3)}\quad \tilde{L}_X = D_X^{-1/2}\,(U_1 \ \ldots \ U_{d^*}) \quad \text{and}$$

$$\tilde{S}_X = (X - 1_n m_X{}')\tilde{L}_X\ .$$

Calculate the QR–decomposition of $\tilde{S}_X$ as

(A.4.4) $\tilde{S}_X = Q\,R$, with
orthogonal $n \times d^*$ matrix Q and upper triangular $d^* \times d^*$ matrix R .

Let the $d^* \times d^*$ matrix R^{-1} be the inverse of the matrix R.
The loadings and eigenvalues are defined as

$$\text{(A.4.5)}\quad L_X = \tilde{L}_X R^{-1}\ ,$$

$$L_Y = D_Y^{-1/2}(V_1 \ \ldots \ V_{d^*})\ ,$$

$$c = (S_1 \ \ldots \ S_{d^*})\ .$$

A.5 Canonical Correlation Analysis CCA

CCA

INPUT : X, Y, d ;

OUTPUT : L_X, L_Y, c .

Let X be an $n \times p$ matrix, Y an $n \times q$ matrix of real values. Let the dimension d be a positive integer with $d \leq \min(p, q, (n-1))$.

A canonical correlation analysis (CCA) between X and Y gives loadings and canonical correlations. The loadings are a $p \times d^*$ matrix L_X and a $q \times d^*$ matrix L_Y, with $d^* \leq d$ and the canonical correlations are a d^*–dimensional vector c . Loadings and canonical correlations are computed as follows.

Let m_X and m_Y be the means of the matrices X and Y .

The truncated singular value decomposition of the centered data then gives

(A.5.1) $(X - 1_n m_X') \approx U_X S_X V_X'$, with
$n \times r_X$ matrix U_X, $p \times r_X$ matrix V_X and diagonal $r_X \times r_X$ matrix S_X,

and

(A.5.2) $(Y - 1_n m_Y') \approx U_Y S_Y V_Y'$, with
$n \times r_Y$ matrix U_Y, $q \times r_Y$ matrix V_Y and diagonal $r_Y \times r_Y$ matrix S_Y.

r_X and r_Y are the maximal integers, such that the diagonal elements of S_X are all greater than εS_{X11} and the diagonal elements of S_Y are all greater than εS_{Y11} , respectively; if either S_{X11} or S_{Y11} are 0 , then CCA cannot be applied.

The truncated singular value decomposition of $U_X' U_Y$ gives

(A.5.3) $U_X' U_Y \approx USV'$, with
$r_X \times d^*$ matrix U, $r_Y \times d^*$ matrix V and diagonal $d^* \times d^*$ matrix S , where $d^* = \min(d, r_X, r_Y, r_0)$.

r_0 is the maximal integer such that the diagonal elements of S are all greater than $\varepsilon\, S_{11}$; if S_{11} is 0 , then r_0 is set to 1.

Let D_X be the diagonal $p \times p$ matrix of variances of X; then the loadings and canonical correlations are defined as

(A.5.4) $L_X = V_X S_X^{-1} U$,
$L_Y = V_Y S_Y^{-1} V$,
$c = \text{diag}(S)$.

A.6 Reduced Rank Regression with Diagonal Error Covariance Matrix RRR

RRR

INPUT : X, Y, d ;

OUTPUT : L_X , c .

Let X be an $n \times p$ matrix, Y an $n \times q$ matrix of real values. Let the dimension d be a positive integer with $d \leq \min(p, q, (n - 1))$.

A reduced rank regression with diagonal error covariance matrix (RRR) between X and Y gives loadings and eigenvalues. The loadings are a $p \times d^*$ matrix L_X , with $d^* \leq d$; the eigenvalues are a d^*–dimensional vector c . Loadings and eigenvalues are computed as follows. Let m_X and m_Y be the means of the matrices X and Y , and define the diagonal $q \times q$ matrix D_Y as the variance of the matrix Y.

The truncated singular value decomposition of the centered matrix X then gives

(A.6.1) $(X - 1_n m_X') \approx U_X S_X V_X'$, with
$n \times r_X$ matrix U_X, $p \times r_X$ matrix V_X and diagonal $r_X \times r_X$ matrix S_X.

r_X is the maximal integer, such that the diagonal elements of S_X are all greater than εS_{X11}; if S_{X11} is 0 , then RRR cannot be applied. Define the $r_X \times q$ matrix

(A.6.2) $C = U_X'(Y - 1_n m_Y')$.

Then initialize $d^* = r_X$ and $S_Z = D_Y$ and, until convergence, iterate as follows.

Calculate the truncated singular value decomposition of $CS_Z^{-1/2}$ as

(A.6.3a) $CS_Z^{-1/2} \approx USV'$, with
$r_X \times d^*$ matrix U, $q \times d^*$ matrix V and diagonal
$d^* \times d^*$ matrix S , where $d^* = \min(d, r_X, r_0)$.

r_0 is the maximal integer such that the diagonal elements of S are all greater than $\varepsilon\, S_{11}$; if S_{11} is 0 , then r_0 is set to 1 and the iteration ends.

Define the $n \times d^*$ matrix S_C and the $n \times q$ matrix Z as

(A.6.3b) $S_C = U_X U$ and $Z = Y - S_C(S_C'Y)$.

Then redefine

(A.6.3c) $S_Z = \mathrm{Diag}(\,\mathrm{diag}(\,Z'Z)\,)$.

Convergence is checked as follows. Define at the i–th iteration $c_i = \mathrm{diag}(S)$. Then if $\| c_i - c_{i-1} \| < \varepsilon$, convergence has been achieved.

After convergence, define the loadings and eigenvalues as

(A.6.4) $L_X = V_X S_X^{-1} U$,
$c = \mathrm{diag}(S)$.

A.7 Redundancy Analysis RDA

RDA

INPUT : X , Y , d , v_Y ;

OUTPUT : L_X , L_Y , c .

Let X be an $n \times p$ matrix, Y an $n \times q$ matrix of real values. Let the dimension d be a positive integer with $d \le \min(p, q, (n-1))$. Let v_Y be an indicator variable, which is equal to 1 if the Y matrix is to be scaled (RDA–COR) and equal to 0 otherwise (RDA–COV).

A redundancy analysis (RDA) between X and Y gives loadings and redundancies. The loadings are a $p \times d^*$ matrix L_X and a $q \times d^*$ matrix L_Y , with $d^* \le d$ and the redundancies are a d^*–dimensional vector c . Loadings and redundancies are computed as follows.

Let m_X and m_Y be the means of the matrices X and Y.

If $v_Y = 1$, define the diagonal $q \times q$ matrix D_Y as the variance of the matrix Y;

if $v_Y = 0$, define D_Y as the $q \times q$ identity matrix I_q.

The truncated singular value decomposition of the centered matrix X then gives

(A.7.1) $(X - 1_n m_X') \approx U_X S_X V_X'$, with
$n \times r_X$ matrix U_X, $p \times r_X$ matrix V_X and diagonal $r_X \times r_X$ matrix S_X.

r_X is the maximal integer, such that the diagonal elements of S_X are all greater than $\varepsilon\, S_{X11}$; if S_{X11} is 0, then RDA cannot be applied.

The truncated singular value decomposition of $U_X'(Y - 1_n m_Y')D_Y^{-1/2}$ gives

(A.7.2) $U_X'(Y - 1_n m_Y')D_Y^{-1/2} \approx USV'$, with
$r_X \times d^*$ matrix U, $q \times d^*$ matrix V and diagonal $d^* \times d^*$ matrix S, where $d^* = \min(d, r_X, r_0)$.

r_0 is the maximal integer such that the diagonal elements of S are all greater than $\varepsilon\, S_{11}$; if S_{11} is 0, then r_0 is set to 1.

Let D_X be the diagonal $p \times p$ matrix of variances of X; then the loadings and redundancies are defined as

(A.7.3) $L_X = V_X S_X^{-1} U$,
$L_Y = D_Y^{-1/2} V$,
$c = \text{diag}(S)$.

A.8 Software

The algorithms specified above are easily programmed in any matrix language such as the S–Plus software (Statistical Science Inc., 1993) or SAS/IML (SAS Institute Inc., 1990). For the estimation of the statistical methods described in chapters 2 to 4, FORTRAN programs as implemented in the OMEGA software have been used (see Schmidli and Weihs, 1994). For the methods described in chapter 5, both FORTRAN programs and functions written in S–Plus have been used.

A.9 Matrix Algebra Definitions

An n–dimensional vector x with elements $x_1, ..., x_n$ is written as

$$x = (x_1, ..., x_n) .$$

Generally, small letters denote vectors or scalars. Special vectors are the n–dimensional vector of 1's, defined as

$$1_n = (11...1)' ,$$

and the n–dimensional vector of 0's, defined as

$$0_n = (00...0)' .$$

An $n \times p$ matrix X with element x_{ij} on row i and column j is written as

$$X = (x_{ij}) .$$

Large letters usually denote matrices. The $n \times n$ unit or identity matrix with 1's on its diagonal and 0's otherwise is denoted by I_n . The elements of the $n \times p$ null matrix $0_{n \times p}$ are all 0 . $0_{p \times p}$ is usually written as 0_p .

For any square $n \times n$ matrix X , x = diag(X) is the n–dimensional vector consisting of the diagonal elements of X . For any n–dimensional vector x , X = Diag(x) is a diagonal $n \times n$ matrix with the elements of x on it's diagonal. For any $n \times p$ matrix X , the vectorization of X is the np–dimensional vector vec(X) obtained by piling the columns of X above each other.

For an $n \times p$ matrix X and an $m \times q$ matrix Y , the Kronecker product of X and Y is the $nm \times pq$ matrix

$$X \otimes Y = \begin{pmatrix} x_{11}Y & \cdots & x_{1p}Y \\ \vdots & & \vdots \\ x_{n1}Y & \cdots & x_{np}Y \end{pmatrix}$$

If X is an $n \times p$ matrix with $X'X = I_n$, then X is called column orthogonal; if X is a square column orthogonal matrix, then X is called orthogonal.

The singular value decomposition (SVD) of an $n \times p$ matrix X is a decomposition of the matrix X into three matrices U, S , V ,

$X = USV'$, with
$n \times p$ column orthogonal matrix U , $p \times p$ orthogonal matrix V and diagonal $p \times p$ matrix S , where the diagonal elements of S are non–negative and ordered by size (smallest element last).

The QR decomposition of an $n \times p$ matrix X is a decomposition of the matrix X into two matrices Q, R,
$X = QR'$, with
$n \times p$ column orthogonal matrix Q , $p \times p$ matrix R with 0's below the diagonal.

Let X be a symmetric, positive definite $p \times p$ matrix, then it's square–root $X^{1/2}$and the inverse of the square root $X^{-1/2}$ are defined as

$X^{1/2} = VS^{1/2}V'$ and
$X^{-1/2} = VS^{-1/2}V'$, where
$X = VSV'$ is the singular value decomposition of X .

For any quadratic $n \times n$ matrix X, it's trace $\mathrm{tr}\, X$ is defined as the sum of its diagonal elements. The Frobenius norm for any $n \times p$ matrix X, is defined by

$$\| X \|^2 = \mathrm{tr} \;\; X'X .$$

Let X be a symmetric, positive definite $p \times p$ matrix, then it's determinant $| X |$ is the product of the diagonal elements of the diagonal matrix S, where $X = VSV'$ is the singular value decomposition of X.

A.10 Multivariate Distributions

A p–dimensional random vector x has a multivariate normal distribution $N(\mu, \Sigma)$ with p–dimensional mean vector μ and $p \times p$ covariance matrix Σ , if its density is given by

$$\rho(x) = \frac{1}{\sqrt{2\pi|\Sigma|}} \exp\{ -\frac{1}{2} (x - \mu)'\Sigma^{-1}(x - \mu) \} .$$

An $n \times p$ random matrix X has a matrix normal distribution $N(M, \Sigma)$ with $n \times p$ mean matrix M and $np \times np$ covariance matrix Σ , if $\mathrm{vec}(X)$ has a multivariate normal distribution with mean vector $\mathrm{vec}(M)$ and covariance matrix Σ .

For any sufficiently regular surface $\Phi \subset R^{n \times p}$, the constraint normal distribution $N_{\Phi}(M, \Sigma)$ with $n \times p$ mean matrix M and $np \times np$

covariance matrix Σ, denotes a distribution on Φ. Its density on the surface Φ is defined to be proportional to the normal density $N(M, \Sigma)$.

If the $n \times p$ random matrix X has a matrix normal distribution $N(0_{n \times p}, \Sigma \otimes I_n)$ with $p \times p$ covariance matrix Σ, then the distribution of $X'X$ is called a Wishart distribution $W(n, n\Sigma)$ with n degrees of freedom. If S has a Wishart distribution $W(n, n\Sigma)$, then the distribution of S^{-1} is called an inverseWishart distribution $W^{-1}(n, n\Sigma)$.

References

Ahn, S.K., Reinsel, G.C. (1988) : Nested Reduced Rank Autoregressive Models for Multiple Time Series; Journal of the American Statistical Association, 83, 849–856.

Aitchison, J., Dunsmore, I. (1975) : Statistical Prediction Analysis; University Press, Cambridge, England.

Akaike, H. (1974) : A New Look at the Statistical Model Identification; IEEE Transactions on Automatation and Control, AC–19m, 716–723.

Akaike, H. (1985) : Prediction and Entropy; in Atkinson, A.C. and Fienberg, S.E. (eds.) : A Celebration of Statistics, 1–24, Springer, New York.

Allen, D.M. (1971) : Mean Squared Error of Prediction as a Criterion for Selecting Variables; Technometrics, 13, 3, 469–475.

Allen, D.M. (1974) : The Relationship between Variable Selection and Data Augmentation and a Method of Prediction; Technometrics, 16, 1, 125–127.

Anderson, T.W. (1951) : Estimating Linear Restrictions on Regression Coefficients for Multivariate Normal Distributions; Annals of Mathematical Statistics, 22, 327–351.

Belsley, D.A., Kuh, E., Welsh, R.E. (1980) : Regression Diagnostics, Wiley, New York.

Besag, J. (1974) : Spatial Interaction and the Statistical Analysis of Lattice Systems (with Discussion); Journal of the Royal Statistical Society B, 36, 192–326.

Borth, D.M., McKay, R.J., Elliott, J.R. (1985) : A Difficulty Information Approach to Substituent Selection in QSAR Studies; Technometrics, 27, 1, 25–35.

Box, G.E.P., Tiao, G.C. (1973) : Bayesian Inference in Statistical Analysis; Addison–Wesley, Reading.

Braak, C.J.F. ter (1990) : Interpreting Canonical Correlation Analysis through Biplots of Structure Correlations and Weights; Psychometrika, 55, 3, 519–531.

Breiman, L., Friedman, J.H. (1985) : Estimating Optimal Transformations for Multiple Regression and Correlation (with Discussion); Journal of the American Statistical Association, 80, 580–619.

Breiman, L., Friedman, J.H. (1994) : Predicting Multivariate Responses in Multiple Linear Regression; Department of Statistics, University of California, Berkley, and Stanford University, Stanford, CA USA.

Breiman, L., Spector, P. (1992) : Submodel Selection and Evaluation in Regression. The X–Random case; International Statistical Review, 60, 291–319.

Bunke, O., Droge, B. (1984) : Bootstrap and Crossvalidation Estimates of the Prediction Error for Linear Regression Models; Annals of Statistics, 12, 4, 1400–1424.

Cook, R.D., Weisberg, S. (1982) : Residuals and Influence in Regression; Chapman and Hall, New York.

Cox, D.R., Snell, E.J. (1974) : The Choice of Variables in Observational Studies; Applied Statistics, 23, 1, 51–59.

Davies, P.T. , Tso, M.K.–S. (1982) : Procedures for Reduced Rank Regression; Applied Statistics, 31, 3, 244–255.

Dawid, A.P. (1993) : Taking Prediction Seriously; manuscript, Dept. of Statistical Science, University College, London.

Dempster, H.P. , Laird, N. , Rubin, D.B. (1977) : Maximum Likelihood from Incomplete Data via the EM Algorithm (with Discussion); Journal of the Royal Statistical Society B, 39, 1–38.

Dewar, M.J.S. , Thiel, W. (1977) : Journal of the American Chemical Society, 99, 4899, 4907, 5231.

Dunn III,W.J. (1989) : Quantitative Structure–Activity Relationships (QSAR); Chemometrics and Intelligent Laboratory Systems, 6, 181–198.

Efron, B. (1983) : Estimating the Error Rate of a Prediction Rule : Improvement on Crossvalidation; Journal of the American Statistical Association, 78, 316–331.

Frank, I.E. (1991) : Variations on a Theme : Partial Least Squares, Intermediate Least Squares and Nonlinear Partial Least Squares; manuscript, Jerll, Inc., Stanford, CA USA.

Frank, I.E., Friedman, J.H. (1993) : A Statistical View of some Chemometrics Regression Tools (with Discussion); Technometrics, 35, 2, 109–148.

Friedman, J.H. (1985) : Classification and Multiple Regression through Projection Pursuit; Technical report 12, Department of Statistics, Stanford University, Stanford, CA USA.

Geisser, S. (1975) : The Predictive Sample Reuse Method with Applications; Journal of the American Statistical Association, 70, 320–328.

Geisser, S. (1980) : A Predictivistic Primer; in Zellner, A. (ed.) : Bayesian Analysis in Econometrics and Statistics, North–Holland, Amsterdam, 363–381.

Geisser, S., Eddy, W.F. (1979) : A Predictive Approach to Model Selection; Journal of the American Statistical Association, 74, 153–160.

Gelfand, A.E. , Dey D.K. , Chang, H. (1992) : Model Determination Using Predictive Distributions with Implementation via Sampling–Based Methods (with Discussion); in Bernardo, J.M., Berger, J.O., Dawid, A.P., Smith, A.F.M. (eds.) : Bayesian Statistics 4; University Press, Oxford, 147–167.

Gelfand, A.E. , Hills, S.E. , Racine–Poon, A. , Smith, A.F.M. (1990) : Illustration of Bayesian Inference in Normal Data Models Using Gibbs Sampling; Journal of the American Statistical Association, 85, 972–985.

Gelfand, A.E. , Smith, A.F.M. (1990) : Sampling–Based Approaches to Calculating Marginal Densities; Journal of the American Statistical Association, 85, 398–406.

Gelman, A., Rubin, D.B. (1992) : Inference from Iterative Simulation using Multiple Sequences; Statistical Science, 7, 457–472.

Geweke, J. (1993) : Bayesian Reduced Rank Regression in Econometrics; manuscript, Federal Reserve Bank of Minneapolis, USA.

Gifi, A. (1990) : Nonlinear Multivariate Analysis; Wiley, New York.

Gilks, W.R. (1993) : Markov Chain Monte Carlo – a Guide for Practice; manuscript, Medical Research Council Biostatistics Unit, Cambridge, UK.

Glasbey, C.A. (1992) : A Reduced Rank Regression Model for Local Variation in Solar Radiation; Applied Statistics, 41, 2, 381–387.

Gnanadesikan, R. (1977) : Methods for Statistical Data Analysis of Multivariate Observations; Wiley, New York.

Gnanadesikan, R., Kettenring, R. (1972) : Robust Estimates, Residuals, and Outlier Detection with Multiresponse Data; Biometrics, 28, 81–124.

Gudmundsson, G. (1977) : Multivariate Analysis of Economic Variables; Applied Statistics, 26, 1, 48–59.

Hansch, C., Leo, A. (1979) : Substituent Constants for Correlation Analysis in Chemistry and Biology; Wiley, New York.

Helland, I.S. (1990) : Partial Least Squares Regression and Statistical Models; Scandinavian Journal of Statistics, 17, 97–114.

Helland, I.S. (1992) : Maximum Likelihood Regression on Relevant Components; Journal of the Royal Statistical Society B, 54, 2, 637–647.

Höskuldsson, A. (1988) : PLS Regression Methods; Journal of Chemometrics, 2, 211–228.

Israëls, A.Z. (1984) : Redundancy Analysis for Qualitative Variables; Psychometrika, 49, 331–346.

Izenman, A.J. (1975) : Reduced Rank Regression for the Multivariate Normal Model; Journal of Multivariate Analysis, 5, 248–264.

Izenman, A.J. (1986) : Reduced Rank Regression Procedures for Discriminant Analysis; American Statistical Association Proceedings of the Statistical Computing Section, 249–253.

Jöreskog, K.G. , Goldberger, A.S. (1975) : Estimation of a Model with Multiple Indicators and Multiple Causes of a Single Latent Variable; Journal of the American Statistical Association, 70, 631–639.

Jöreskog, K.G. (1967) : Some Contributions to Maximum Likelihood Factor Analysis; Psychometrics, 32, 443–482.

Judge, G.G., Griffiths, W.E., Hill, R.C., Lütkepohl, H., Lee, T.–C. (1985) : The Theory and Practice of Econometrics; Wiley, New York.

Kaiser, H.F. (1958) : The Varimax Criterion for Analytic Rotation in Factor Analysis; Psychometrika, 23, 187–200.

Leeden, R. van der (1990) : Reduced Rank Regression with Structured Residuals; Leiden, DSWO Press.

Lindley, D.V. (1984) : discussion in Miller (1984).

Lindley, D.V., Smith, A.F.M. (1972) : Bayes Estimates for the Linear Model (with Discussion); Journal of the Royal Statistical Society B, 34, 1–40.

Mager, P.P. (1984) : Multidimensional Pharmacochemistry; Academic Press, London.

Magnus, J.R. , Neudecker, H. (1988) : Matrix Differential Calculus; Wiley, Chichester.

Mallows, C.L. (1973) : Some Comments on Cp; Technometrics; 15, 4, 661–675.

Mardia, K.V., Kent, J.T., Bibby, J.M. (1979) : Multivariate Analysis; Academic Press, London.

Massy, W.F. (1965) : Principal Component Regression in Exploratory Statistical Research; Journal of the American Statistical Association, 60, 234–246.

Meng, X.–L., Rubin, D.A. (1993) : Maximum Likelihood Estimation via the ECM Algorithm : A General Framework; Biometrika, 80, 2, 267–278.

Miller, A.J. (1984) : Selection of Subsets of Regression Variables; Journal of the Royal Statistical Society A, 147, 3, 389–425.

Moser, P., Sallmann, A. , Wiesenberg, I. (1990) : Synthesis and Quantitative Structure–Activity Relationships of Diclofenac Analogs, Journal of Medical Chemistry, 33, 2358.

Nirmalakhandan, N., Speece, R.E. (1988) : Structure–Activity Relationships; Environmental Scientific Technology, 22, 6, 606–615.

Odell, P.L., Feiveson, A.H. (1966) : A Numerical Procedure to Generate a Sample Covariance Matrix; Journal of the American Statistical Association, 61, 198–203.

Pascual, A., Rindlisbacher, A., Schmidli, H., Stamm, E. (1994) : N–(Pyrid–3–yl)thioureas and Derivatives as Acaricides. II. Structure–Activity Relationships and Chemodynamic Behaviour; manuscript, CIBA–GEIGY Ltd., Basel; to appear in Pesticide Science.

Polasek, W. (1988) : Explorative Daten–Analyse; Springer, Heidelberg.

Potthoff, R.F. , Roy, S.N. (1964) : A Generalized Multivariate Analysis of Variance Model useful especially for Growth Curve Models; Biometrika, 51, 313–326.

Roecker, E.B. (1991) : Prediction Error and its Estimation for Subset–Selected Models; Technometrics, 33, 4, 459–468.

Rose, V.S., Hyde, R.M., Macfie, H.J.H. (1990): U.K.Usage of Chemometrics and Artificial Intelligence in QSAR Analysis; Journal of Chemometrics, 4, 355–360.

Ryan, D.A.J. , Hubert, J.J., Carter, E.M. , Sprague, J.B., Parrott, J. (1992) : A Reduced Rank Multivariate Regression Approach to Aquatic Joint Toxicity Experiments; Biometrics, 48, 155–162.

Schervish, M.J. (1989) : Multivariate Analysis; Lecture Notes presented at the Institute of Statistics and Econometrics, University of Basel.

Schmidli, H. (1991) : Ein Konzept für die Analyse multivariater Daten; Technical Report No. 9105; Mathematical Applications, CIBA–GEIGY Ltd., Basel.

Schmidli, H. (1992) : Selection of Multivariate Prediction Methods; Technical Report No. 9206; Mathematical Applications, CIBA–GEIGY Ltd., Basel.

Schmidli, H. (1994) : Bayesian Analysis of Reduced Rank Regression; manuscript; Mathematical Applications, CIBA–GEIGY Ltd., Basel; submitted for publication.

Schmidli, H., Weihs, C. (1994) : OMEGA – User Interface and Algorithms; manuscript; Mathematical Applications, CIBA–GEIGY Ltd., Basel.

Seber, G.A.F. (1984) : Multivariate Observations; Wiley, New York.

Shao, J. (1993) : Linear Model Selection by Crossvalidation; Journal of the American Statistical Association, 88, 486–494.

Sibson, R. (1978) : Studies in the Robustness of Multidimensional Scaling : Procrustes Statistics; Journal of the Royal Statistical Society B, 40, 234–238.

Smith, A.F.M., Gelfand, A.E. (1992) : Bayesian Statistics Without Tears : A Sampling–Resampling Perspective; The American Statistician, 46, 2, 84–88.

Stone, M. (1974) : Cross–Validatory Choice and Assessment of Statistical Predictions (with Discussion); Journal of the Royal Statistical Society B, 36, 111–147.

Stone, M. (1977) : An Asymtotic Equivalence of Choice of Model by Crossvalidation and Akaike's Criterion; Journal of the Royal Statistical Society B, 39, 44–47.

Stone, M. , Brooks, R.J. (1992) : Continuum Regression : Cross–validated sequentially Constructed Prediction Embracing Ordinary Least Squares, Partial Least Squares and Principal Components Regression; Journal of the Royal Statistical Society B, 52, 2, 237–269.

Stone, M. , Jonathan, P. (1993) : Statistical Thinking and Technique for QSAR and Related Studies (Part I : General Theory); Journal of Chemometrics, 7, 455–475.

Stone, M. , Jonathan, P. (1994) : Statistical Thinking and Technique for QSAR and Related Studies (Part II : Specific Methods); Journal of Chemometrics, 8, 1–20.

Stuetzle, W. (1988) : Plot Windows; in Cleveland, W.S., McGill, M.E. (eds.) : Dynamic Graphics for Statistics, Wadsworth, Belmont, 225–245.

Sundberg, R. (1993) : Continuum Regression and Ridge Regression; Journal of the Royal Statistical Society B, 55, 3, 653–649.

Thisted, R.A. (1988) : Elements of Statistical Computing; Chapman and Hall, New York.

Toutenburg, H. (1992) : Lineare Modelle; Physica Verlag, Heidelberg.

Toutenburg, H., Trenkler, G. (1990) : Mean Square Error Matrix Comparisons of Optimal and Classical Predictors and Estimators; Computational Statistics & Data Analysis, 10, 297–305.

Tso, M.K.–S. (1981) : Reduced Rank Regression and Canonical Analysis; Journal of the Royal Statistical Society, 43, 2, 183–189.

Velu, R.P., Reinsel, G.C., Wichern, D.W. (1986) : Reduced Rank Regression Models for Multiple Time Series; Biometrika, 73, 105–118.

Velu, R.P. (1991) : Reduced Rank Models with Two Sets of Regressors; Applied Statistics, 40, 1, 159–170.

Verloop, A., Hoogenstraaten, W., Tipker, J. (1976) : Development and Application of new Steric Substituent Parameters in Drug Design; in Drug Design (Ed.Ariëns, E.J.), Vol.VII (Series in medicinal chemistry, Vol.11–VII), Academic Press, New York, 165–207.

Waterbeemd, H. van de, Testa, B. (1987) : The Parametrization of Lipophilicity and other Structural Properties in Drug Design; Advances in Drug Research, 16, 85–225.

Weihs, C. (1993) : Multivariate Exploratory Data Analysis and Graphics : a Tutorial; Journal of Chemometrics, 7, 305–340.

Weihs, C., Schmidli, H. (1990) : OMEGA (Online Multivariate Exploratory Graphical Analysis): Routine Searching for Structure (with Discussion); Statistical Science, 5, 175–226.

Weihs, C., Schmidli, H. (1991) : Multivariate Exploratory Data Analysis in Chemical Industry; Mikrochimica Acta II, 467–482.

Wold, H. (1982) : Soft Modeling, the Basic Design and some Extensions; in Jöreskog, K.G., Wold, H. (eds.) : Systems under Indirect Observation, North Holland, Amsterdam.

Wold, S. (1977) : Cross–Validatory Estimation of the Number of Components in Factor and Principal Components Models; Technometrics, 20, 4, 397–405.

Wold, S., Sjoestroem, M., Carlson, R., Lundstedt, T., Hellberg, S., Skagerberg, B., Wikstroem, C., Oehman, J. (1986) : Multivariate Design; Analytica Chimica Acta, 191, 17–32.

Wollenberg, A.L. van den (1977) : Redundancy Analysis, an Alternative for Canonical Correlation Analysis; Psychometrika , 42, 207–219.

Main Notations and Abbreviations

ASY	asymptotic (index)
CCA	canonical correlation analysis (reduced rank regression with unstructured error covariance matrix)
CCAv	canonical correlation analysis combined with variable selection
CDA	canonical discriminant analysis
CV	crossvalidation (index)
FIT	fitting (index)
IC	information criterion
i.i.d.	independent and identically distributed
MR	multivariate regression
MSEP	mean squared error of prediction
PCA	principal component analysis
PCR	principal component regression
PLS	partial least squares
QSAR	quantitative structure–activity relationships
RDA	redundancy analysis (reduced rank regression with error covariance matrix proportional to the identity matrix)
RDAv	redundancy analysis combined with variable selection
RRR	reduced rank regression with diagonal error covariance matrix
RRRv	reduced rank regression combined with variable selection

R^2	R–squared value (goodness of fit)
$N(\mu, \Sigma)$	multivariate normal distribution
$N(M, \Sigma)$	matrix normal distribution
$N_{\Phi}(M, \Sigma)$	constraint matrix normal distribution
$W^{-1}(n, \Sigma)$	inverse Wishart distribution
I_p	$p \times p$ identity matrix
1_p	p–dimensional vector of 1's
0_p	$p \times p$ matrix of 0's
$0_{p \times q}$	$p \times q$ matrix of 0's , i.e. $0_{p \times p} \equiv 0_p$
vec(A)	vectorization of a matrix A
tr(A)	trace of a square matrix A
diag(A)	vector consisting of the diagonal elements of the square matrix A
Diag(a)	diagonal matrix whose diagonal elements consist of the elements of the vector a
$A \otimes B$	Kronecker product of the matrices A and B
A_{ij}	element on the i–th row and the j–th column of the matrix A
$A_{i.}$	i–th row of the matrix A
$A_{(i)}$	matrix obtained from the matrix A by the elimination of the i–th row
[A]	distribution of the random matrix A
[A | B]	conditional distribution of the random matrix A given random matrix B

Subject Index

GPSR Compliance

The European Union's (EU) General Product Safety Regulation (GPSR) is a set of rules that requires consumer products to be safe and our obligations to ensure this.

If you have any concerns about our products, you can contact us on ProductSafety@springernature.com

In case Publisher is established outside the EU, the EU authorized representative is:

Springer Nature Customer Service Center GmbH
Europaplatz 3
69115 Heidelberg, Germany

Batch number: 10732714

Printed by Printforce, the Netherlands